P9-APM-875

EINSTEIN'S TWIN

To Marianne,
My Lockdown Muse,
With Love

Copyright © Elwin Street Productions Limited 2020

Conceived and produced by
Elwin Street Productions Limited
14 Clerkenwell Green
London EC1R 0DP
www.modern-books.com

BLOOMSBURY PUBLISHING
Bloomsbury Publishing Inc.
1385 Broadway, New York, NY 10018, USA

BLOOMSBURY, BLOOMSBURY PUBLISHING and the Diana
logo are trademarks of Bloomsbury Publishing Plc

First published in the United States 2020

ISBN: HB: 978-1-63557-586-6

Library of Congress Cataloging-in-Publication Data has been
applied for.

2 4 6 8 10 9 7 5 3 1

Printed in China

EINSTEIN'S TWIN

MIND-BENDING PUZZLES AND PARADOXES

FROM THE WORLD OF SCIENCE

JEREMY STANGROOM

BLOOMSBURY PUBLISHING

NEW YORK • LONDON • OXFORD • NEW DELHI • SYDNEY

Contents

Introduction

I.
CLASSIC PUZZLES AND CONUNDRUMS 8
Married Or Not? 10
Are You The New Sherlock Holmes? 12
Switch Or Stick? 14
How Wide Is The Lake? 16
How Many Cookies Will Be Left? 18
Who Lost What? 20
How Old Is He? 22

2.
PUZZLING PROBABILITIES 24
Who Will Be Executed, Who Will Be Pardoned? 26
How Long Will Their Relationship Last? 28
Are We Living In A Simulation? 30
Which Die Should She Choose? 32
How Many Possible Worlds? 34
What's The Real Benefit? 36
Whose Birthday Is It Anyway? 38

3.
SLIPPERY SCIENCE 40
Is the Cat Dead Or Alive? 42
How Old Is Einstein's Twin? 44
When Is A Wave Not A Wave? 46
Are There Spooky Actions At A Distance? 49
What Did the Telescope Reveal? 52

4.

PERPLEXINGLY PARADOXICAL 54

What Will The Crocodile Do? 56
To Vote Or Not To Vote? 58
The Same Or Not? 60
Will the Tortoise Start The Race? 62
What Happens Next? 64

5.

LIFE, THE EARTH AND EVERYTHING 66

How Do Bees Fly? 68
Over The Limit? 70
Will She Be Faster? 72
Who Wrote The Music? 74
Where Are All The Aliens? 76

Solutions and Discussion 78

Index 142

INTRODUCTION

In September 2011, particle physicists at CERN shocked the world of science by announcing results that suggested subatomic particles called neutrinos are able to travel faster than the speed of light, the universe's inviolable speed limit. This was a huge deal, not only because faster-than-light particles contradict nearly a century's worth of scientific research, but also because they open up the possibility of communicating with the past, which plunges us straight into a world of paradox and contradiction.

Not surprisingly, physicists reacted to the news with skepticism and amazement, with one declaring that he would eat his boxer shorts live on television if the CERN results turned out to be true. Modern physics relies on the truth of quantum mechanics and Einstein's theory of relativity. If it were necessary to give up relativity, which rules out faster-than-light travel, then deep trouble would result. The idea of causality, for example, would be undermined, because the ability to communicate with the past would allow past, present and future to interact with each other, giving rise to conundrums such as the Grandfather Paradox (which we discuss later in the book).

The beauty of science is that it is self-correcting, and, as it turned out, the CERN results were wrong, the error being caused by a loose cable. However, though physics survived its brief wobble, and the threat posed by the possibility of faster-than-light travel retreated into the background, it would be wrong to think that scientific orthodoxy is ever comfortable. There remain puzzles and paradoxes that are yet to be solved, and gaps in our knowledge that currently appear unbridgeable.

As you work through the chapters in this book, you're going to meet science's biggest puzzles and mysteries, and confront some of its most esoteric ideas, including time dilation, wave-particle duality and quantum entanglement. The Nobel laureate Philip Anderson once noted that science renews itself through a seemingly endless supply of new questions, inspired by the answers to old questions. The hope is that by the time you finish this book, you'll have a strong sense of the challenges that drive science at its frontiers, and also an awareness of some of the fundamental laws that underpin the workings of the universe.

1
CLASSIC PUZZLES AND CONUNDRUMS

*Contrariwise...if it was so, it might be;
and if it were so, it would be; but as it isn't, it ain't. That's logic*
LEWIS CARROLL (*THROUGH THE LOOKING GLASS*)

Lewis Carroll once instructed that we should begin at the beginning and go on until we come to the end. Happily, though we must defer to Carroll in matters logical, there is no need to approach this book in that fashion. You attempt the puzzles in any order that takes your fancy.

The puzzles in this chapter are not easy exactly, but they all have solutions. If you figure out the logic, which, of course, is easier said than done, then you should be able to get to the right answer.

These puzzles are all classics of their type. To solve them, you'll need a mixture of logic, lateral thinking, prowess with probability, and easy math. Don't worry if you find one or two of them discombobulating and beyond your powers to solve. Part of the reason the puzzles here have stood the test of time is precisely that they're a little bit tricky. But hopefully with minimal gnashing of the teeth and not too many desperate entreaties for help at least some of them will give up their secrets to you.

MARRIED OR NOT?

The villagers of Chudleigh-by-the-Pond have for many years been proud to rely upon the Greater Chudleigh registry office to record their births, marriages and deaths. However, a change in management, new mortality targets, and a series of mishaps have dented their confidence in the probity of that august institution.

Lurid headlines in *Pond Life*, Chudleigh's local newspaper, haven't helped. In fact, it was one such headline

— *DEAD CAT MARRIES WOMAN* —

that prompted the parish council to act in the hope of sorting out the mess at the registry office. Realizing that outside expertise was required, it called in ex-policeman and peripatetic philosopher, Inspector Horse, and asked him to spend a day surveying registry work practices.

Horse reported back that the problem lay in recruitment. The current staff lacked expertise in the areas of birth, marriage, death, and the recording of these things. Luckily, the solution would be clear-cut. Going forward, all staff would have to pass a registry office logic test before being let loose in the records department. The test has the following form.

☞ Bentley, Estella and Philip turn up at the registry office. Bentley is looking at Estella, but Estella is looking at Philip. Bentley is married, but Philip is not.

The question itself is straightforward: Is a married person looking at an unmarried person?

Horse explains that the question involves no trickery or

wordplay designed to throw people off the scent. Nobody is divorced or widowed. The three protagonists are all people, not the cat of the *Pond Life* headline. It really is entirely as straightforward as it seems.

There are three possible answers:

(a) Yes (b) No (c) Cannot be determined. 👉

Do you have what it takes to work at the Chudleigh-by-the-Pond registry office?

(SOLUTION ON PAGE 80)

ARE YOU THE NEW SHERLOCK HOLMES?

Ex-police officer Jack Dawe has been working in investigation and security for a number of years now, keeping the Greater Chudleigh area safe from bludgers, resurrectionists and other miscreants. He has always been content to work on his own, but after recent events he has reluctantly come to the conclusion that he needs to recruit a specialist detective to work under his command.

The theft of Henry the famous sturgeon from Chudleigh's aquarium while Dawe was tasked with the aquarium's security was bad enough, but the fact that Henry was entered into the village's annual fishing contest, and swam off with first prize, has exploded Dawe's reputation into the dust.

Dawe hopes that recruiting new blood will help him restore his reputation to its former glory. But he has to be careful. He doesn't want to recruit a dope to the role; he needs somebody with a razor-sharp logical mind, a philosopher king if you like. So he devises a test of logic to weed out the slapdash and infirm of mind.

☞ The rules are simple: For each of the three arguments below, decide if the given conclusion follows logically from the premises. Answer YES if, and only if, the conclusion can be derived as a matter of certainty from the given premises. Otherwise, answer NO. ☜

	Premise	Premise	Conclusion
Argument 1	All things that are swallowed are good for the health	Alcohol is swallowed	Alcohol is good for the health
Argument 2	Every animal with a tail is vicious	A golden retriever is not vicious	A golden retriever is not an animal with a tail
Argument 3	All short people have orange skin	Ronald Plump is not short	Ronald Plump does not have orange skin

Let's assume you want to score maximum marks so you have the chance to join Jack Dawe in his mission to keep the good people and sturgeons of Chudleigh safe.

What answer should you give for each of these arguments?

(SOLUTION ON PAGE 82)

SWITCH OR STICK?

The villagers of Chudleigh-by-the-Pond are used to the shenanigans of The Three Bears Massive, Chudleigh's most notorious street gang. Tormented tulips, stolen sturgeons, hacked hedgerows — the villagers have seen it all.

Or so they thought. But even the most hardened among them have been shocked by the events currently unfolding at old Barnaby's farm. The Three Bears Massive have strapped a bomb to Lola, a prize-winning and much-loved sheep, and are threatening to turn her into aerial mutton unless their demand for unlimited shearing rights within the boundaries of Greater Chudleigh is met.

Jack Dawe, Chudleigh's much maligned security operative, had been attempting to negotiate Lola's release, but tempted by an offer of a cream tea, he's managed to get himself taken hostage. This explains why he's currently engaged in a bizarre game of chicken, the result of which

will determine the fate of Chudleigh's beloved ungulate.

The Three Bears Massive show Dawe a box with three identical buttons, and explain that two of the buttons are wired up to a timer that will run for five minutes before triggering an explosion that'll blast Lola into the stratosphere. The third button, however, will detach Lola from the explosive device and allow her to escape.

Dawe has no way of telling which is the escape button, so it seems there's a one-in-three chance. However, the furry fiends add in a complication. Dawe is told to choose a button, but not to press it. A gang member, who knows how the box is wired up, will then press one of the two remaining buttons, ensuring that it's a trigger button. Dawe then has a couple of minutes to choose whether to keep to his original choice or to switch and press the other remaining button. If he presses the second trigger button, Lola will die. If he presses the escape button, Lola will live to *baa baa* another day.

Should Dawe stick to his
original button or should he switch?
And what's the rationale for his choice?

(SOLUTION ON PAGE 84)

HOW WIDE IS THE LAKE?

After a series of mishaps off the coast of a channel island, the *Tribulation* ferry boat has been towed to Lake Chudleigh, where it will undergo a series of seaworthiness tests. The intention is to get it shipshape, or at the very least to reduce the amount of listing, stalling and sinking.

The first test is of its speed. But the *Tribulation*'s captain, Nicholas Trink, has a problem. The ship's pitometer and odometer are both broken. He has no way of knowing how fast the ship is moving or how far it has traveled.

Luckily enough, while drowning his sorrows in the Ship Inn, Trink bumps into Rear Admiral Tilly Westry, who has been running maneuvres for the Royal Navy on Lake Chudleigh. She listens to his tale of woe, and then suggests a solution to his problem, one that will allow him to work out how far the *Tribulation* has traveled and by extension, its average speed during the journey. This is her suggestion:

☞ He will set off in the *Tribulation* from one side of Lake Chudleigh at the same time as she sets off in the HMS *Rapide* from the other side of the lake. Both ships will set a constant, though different, speed. As the ships pass each other, the *Rapide* will make use of its sophisticated satellite technology to determine how far the *Tribulation* has traveled from shore. Once the ships reach dry land (on opposite sides of the lake, obviously) they'll pause for 10 minutes, and then return the other way. As the ships pass each other, the *Rapide* will calculate how far the *Tribulation* has traveled from shore on its return journey. The ships will then complete their journeys.

In his drunk state, Captain Trink agrees to this plan. The next day, the *Tribulation* and *Rapide* successfully complete the plan as outlined by Rear Admiral Westry. Captain Trink is informed that the *Tribulation* was 720 yards from its starting point when the ships first passed each other, and it

was 400 yards from the other shore when the ships passed each other in the opposite direction.

But Trink is stumped. He cannot figure out how he's supposed to use this data to calculate how far the *Tribulation* has traveled. Can you help him out?

How wide is the lake? How far did the Tribulation travel?

(SOLUTION ON PAGE 86)

HOW MANY COOKIES ARE LEFT?

Reba French-Davies, a Tibetan terrier of impeccable provenance, has excelled in the first week of her new job as a canine courier, expediting the delivery of packages of all sizes – though small, mainly – throughout the Greater Bovey area.

A pot of honey, beeswax wraps, a ceramic hare, and oven gloves, were all transported in a timely fashion to the happy customers of Whippet Over, Bovey's most successful delivery company staffed only by dogs.

It is unfortunate that Reba's burgeoning reputation as the handiest hound in town is under threat from what seems to be an unlikely source. Cookies. Or, more precisely, 3000 cookies to be delivered to a bakery in John O'Groats, which, as luck would have it, lies exactly 1000 kilometers away from the Bovey headquarters of Whippet Over.

The problem is that Reba really loves cookies, more than her job, more than pussy cats, even. Around a cookie, she has no impulse control. As far as Reba is concerned, a cookie is something to be eaten as fast as possible, ideally in a single mouthful, not something to be transported 1000 kilometers.

Reba is an honorable sort of hound, so she

explains to her boss that really it's very unlikely that 3000 cookies are going to arrive at John O'Groats. She estimates that for every one kilometer she travels, she will snarf a single cookie. She also explains that the maximum number of cookies she can carry at the same time is 1000.

Neither Reba nor her boss are entirely sure what this means in terms of the John O'Groats delivery. Exactly how many cookies are going to be left at the end of her journey?

How many of the 3000 cookies will be left after Reba has traveled the 1000 kilometers to John O'Groats?

(SOLUTION ON PAGE 88)

Tips: (a) Reba can only carry 1000 cookies at a time; (b) you need to identify the optimal strategy for the journey so you can work out the maximum number of cookies that will arrive at John O'Groats.

WHO LOST WHAT?

The Bovey Amateur Dramatics Society is currently attracting stellar reviews for its lavish production of *Les Misérables*. Theatergoers from the four corners of Dartmoor have flocked to Bovey's village hall, enticed partly by silversmith John Penn's postmodern reimagination of the role of Inspector Javert, which involves a series of silences each lasting precisely four minutes and thirty-three seconds, but also by the fact that tickets cost only 1 shilling and 6 pence.

Alex Gibbon, former professor of sociology at North Bovey Institute of Technology, has not been able to see the show. Now an itinerant ukulele virtuoso, he has only 1 shilling to his name, so can't afford to buy a ticket even at 1930s prices. Driven by a burning sense of injustice, plus a great love for "I Dreamed a Dream," Gibbon hits upon a cunning plan to secure himself a ticket.

Before decimalization, the UK used a currency consisting of pounds, shillings and pence.
1 shilling = 12 pence, so the cost of a theater ticket, at 1 shilling and 6 pence, was 18 pence.

☞ Gibbon took his 1 shilling to a pawnbroker. The pawnbroker lent Gibbon 9 pence against the shilling, and handed over a pawn ticket so Gibbon could get his shilling back at a later date. Not a wise deal, you might think, given that Gibbon now has only 9 pence to his name. Except Gibbon sells the pawn ticket to a friend for 9 pence, which means he now has 18 pence – 1 shilling and 6 pence – enough to allow him to see the show. ☜

How has this happened? From where did the extra 6 pence spring? Who has lost what?

(SOLUTION ON PAGE 91)

HOW OLD IS HE?

The Reverend Daniel Martin, vicar of St Equis Parish Church, Greater Bovey, one of Dartmoor's leading centers of ecumenical inquiry, has issued a cunning decree to help prepare his novices for their role as roving outreach apostles: they are not allowed to respond to a straight question with a straight answer.

This has worked well. The novices have become adept at sidestepping such tricky issues as the Plotinian origins of apophatic theology, the character of Christ as Logos, and whether or not God actually exists.

However, right now, Daniel is doubting the wisdom of his decree as he struggles to figure out the age of a new member of his regular seminar on Tim LaHaye and the challenges of eschatology. All he wants to do is find out a little bit about the fellow, somebody with whom he expects to share many profound learning experiences, but nobody is playing ball.

Daniel has managed to figure out by means of a name tag that the new guy goes by the name of Charon, but that's about it. He has asked him his age straight-out, of course, but Charon, backed up by his novice colleagues, refuses to give him a straight answer. It would be against his decree, they all say. Daniel tries to explain that his decree is designed to deal with matters theological, but they just smile at him, and tell him they're wise to his serpent ways and have no intention of failing what is clearly a test of their fidelity.

After an hour or two of theological jousting and custard creams, the novices get together in a huddle. When they emerge, they tell Daniel they'll show him a puzzle that will allow him to deduce the age of Charon.

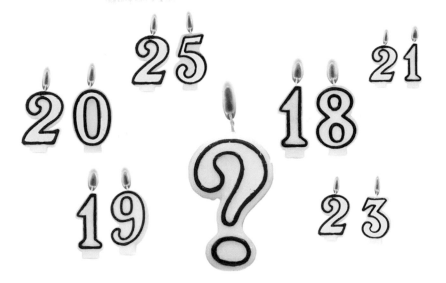

☞ The sum of the ages of Charon and William is 32. In two years' time, Charon will be three times as old as William. How old is Charon now? ☜

The Reverend Daniel immediately objects that he doesn't know who this William person is, but the novices respond that it doesn't matter, he should be able to work out Charon's age anyway.

Can you help the esteemed Reverend by figuring out the age of this fishy fellow Charon?

(SOLUTION ON PAGE 94)

Who Will Be Executed, Who Will Be Pardoned? • How Long Will Their
Relationship Last? • Are We Living In A Simulation? • Which Die Should
She Choose? • How Many Possible Worlds? • What's The Real Benefit?
Whose Birthday Is It Anyway?

2

PUZZLING PROBABILITIES

*When a coincidence seems amazing, that's because the human mind isn't
wired to naturally comprehend probability & statistics.*
NEIL deGRASSE TYSON

How well do you think you'd do if you were asked to
calculate the risk associated with a dangerous activity? A
simple test, first devised by Dan Morris, will perhaps tell us
whether you have an aptitude for these kinds of probability
calculations. Let's suppose you are randomly breathalyzing
drivers to check if they are over the legal drinking limit.

The following things are true: 1) If a driver is drunk,
the breathalyzer will capture this fact with 100% accuracy.
2) 1 in 1000 drivers will be drunk. 3) The probability the
breathalyzer will show a positive result is 4%.

The question is — if a randomly selected driver tests
positive, what's the likelihood they're actually drunk?

Would it surprise you to hear that if a randomly selected
driver fails our breathalyzer test, the probability that they're
actually drunk is only 2.5%?

If not, then there's a good chance you're going to sail
through the probability puzzles that feature in this chapter. If
you are surprised, then buckle up, you're in for a challenging
and eye-opening time of things.

HOW MANY POSSIBLE WORLDS?

Marlon Malloy never imagined that he'd be lying on the couch of Dr Lewis Friend, metaphysician and philosophical counselor, trying to sort his life out. He'd been a top-notch amateur boxer in his youth, but had given it all up for a girl and a life of hard work and drudgery on the docks. Now, as he entered his twilight years, he couldn't escape the feeling that he'd wasted his life. He could have been a contender, he told himself, if only he'd kept up with the boxing. Fame and fortune had beckoned, and instead he'd chosen a broken marriage and a bad back.

At least he'd always believed that he'd made this choice, but Dr Friend has suggested that a different version of himself, a counterpart, had actually chosen the boxing. Nonsense, of course, but Marlon can't quite nail down where it goes wrong.

MARLON: *So you're saying that the only way to make sense of the claim that I could have been a contender is to posit the existence of a possible world in which my counterpart is (or was) a contender?*
DR FRIEND: *Yes, that's it. There is nothing in our world to make a*

claim about what could have happened — that you could have become a contender, for example — true. After all, you never became a contender, so we can't appeal to facts or things in the world.

MARLON: *Sounds like you're saying that because I didn't become a contender, I couldn't have become a contender. There's only one world, and in that world I'm a divorced man with a bad back.*

DR FRIEND: *No, not at all. If I say that you could have been a contender, I'm just saying there's a possible world where you, or at least your counterpart, did become a contender. It's a world very like ours, it's close to ours, but it's a world in which you didn't give up boxing.*

MARLON: *But it doesn't actually exist, does it? It's not real. It's not the same kind of world as our world — it's got to be an imaginary world.*

DR FRIEND: *Why does it have to be imaginary? And what good would it do if it were just imaginary? How can something imaginary act as a truth-maker for the claim that you could have been a contender? I might just as well claim that I could have been a contender!*

MARLON: [Stares blankly.]

DR FRIEND: *The fact of the matter is that possible worlds are useful constructs. They help us to understand perfectly reasonable notions of possibility, necessity and contingency. In other contexts, we have no problem with asserting the existence of abstract constructs — take mathematical sets, for example. So what's the difference?*

Is the idea that there are other possible worlds, similar in kind to our world, so much philosophical garbage?

(SOLUTION ON PAGE 104)

WHAT'S THE REAL BENEFIT?

Hotelier Basil Sinclair has had a rough few years. It turns out that owning a hotel with an infinite number of rooms takes a toll. The breakfast rotation is a nightmare, the cleaning bill ridiculous, and with the rise of a certain internet-based BnB enterprise, it's a struggle to fill all the rooms. Even during Iowa's peak tourist season, you're likely to find an empty room or two, which is a travesty considering Des Moines is only a three-hour drive away.

Basil thought he was coping well, but a trip to see his physician, and a full medical workup, says otherwise. He knew he was in trouble when the doctor suggested he lie down and asked whether he'd brought anybody along for support. All kinds of possibilities flooded through his possibly encephalitic head — dropsy, ague, the French pox, or maybe a combination of all three.

Unfortunately, the diagnosis is every bit as scary as he feared. Dr Rieux, Basil's medical man, informs Basil he is suffering from a severe case of phlogiston poisoning, likely to be the consequence of Basil's habit of smothering his food in spicy sauces and fiery seasoning. Happily, there is a recognized treatment — a course of lavoisian pills, which Dr Rieux recommends Basil starts immediately.

Basil is suspicious. He didn't become the owner of Iowa's largest hotel by popping pills willy-nilly. A vodka a day and weekly trip to a sauna has until now sufficed to keep the doctor away. But Dr Rieux is insistent. He explains that the evidence for the therapeutic effect of lavoisian is conclusive. One large clinical study of more than ten thousand patients, showed a 36% reduction in

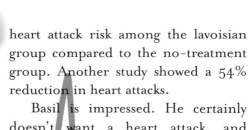

heart attack risk among the lavoisian group compared to the no-treatment group. Another study showed a 54% reduction in heart attacks.

Basil is impressed. He certainly doesn't want a heart attack, and those figures seem persuasive. He asks about side-effects, and Dr Rieux tells him that all medications carry the risk of complications, but lavoisian is a safe drug. He asks whether the clinical studies included patients who like him had never experienced a previous cardiac event, and he's told that both studies included such patients.

This is enough to persuade Basil that lavoisian is going to be part of his future plans. Vodka and saunas can only take a man so far.

But does Basil really have sufficient information to make an informed judgment about lavoisian? Let's assume for the sake of this puzzle that the research studies are methodologically sound and accurately reported. Let's also assume that there are no significant side effects and that lavoisian is safe.

Does Basil know enough to take lavoisian with confidence that its therapeutic effect is significant?

(SOLUTION ON PAGE 106)

WHOSE BIRTHDAY IS IT ANYWAY?

Clare Brogan has enjoyed an uneasy relationship with birthdays ever since her parents threw a surprise party for her 18th birthday in defiance of the laws of good taste and logic. It is a little ironic, then, that she now spends her working life crouched inside a giant cake, out of which she will leap, bringing joyous felicitations to unsuspecting and slightly horrified birthday boys and girls.

It's a highly skilled profession, requiring flexibility, quadriceps, and impeccable timing, but Clare is good at it, and she hasn't once regretted the six months she spent at night school mastering its intricacies and peculiarities. Until now, that is.

She is currently in a little bit of trouble due to a case of mistaken identity. Looking back, perhaps "So who's the birthday boy around here?" hadn't been specific enough an inquiry, and admittedly, it did seem somewhat unlikely that anybody would arrange a pop-up cakeogram for the Archbishop of Canterbury, but how was she supposed to know there were two birthday boys at such a small gathering?

After all, what are the chances? There are only 23 people in the room, so it must be very unlikely that two of them would share the same birthday. Also, it's not as though the archbishop has a twin, nor that there are any twins in the room, so far as Clare can tell.

Unfortunately, Clare's protestations are in vain, and though it's undeniable she looks fine in her steampunk Wonder Woman outfit, she's ejected together with her cake from the birthday soiree (though whose birthday exactly remains something of a mystery).

If Clare is right that it's very unlikely, then maybe she can resume her career in birthday entertainment. Otherwise, it's likely that her days of cake pyrotechnics are over, a sad loss to the pop-up cakeogram industry.

Is Clare right to suppose it's highly unlikely that two people out of a gathering of 23 would share the same birthday?

(SOLUTION ON PAGE 108)

Tips: Firstly, ignore leap years – they just confuse the issue when it comes to this sort of puzzle. Second, there's no trick here. There are 23 people in the room, not including Clare (don't include Clare in your calculations).

3

SLIPPERY SCIENCE

Anyone who is not shocked by quantum theory has not understood it.
NIELS BOHR

The entries featured in this chapter focus on the strange and paradoxical aspects of the quantum world. Ghostly action at a distance, particles that are also waves, cats that are both dead and alive, all make an appearance.

If you're not familiar with quantum mechanics, then you're going to find this chapter challenging. But you should draw some comfort from the thought that you're not alone. Albert Einstein was so discomfited by the implications of quantum mechanics that he spent a good deal of the latter part of his life trying, and failing, to find reasons why it couldn't be true.

There's a famous quote about quantum mechanics, widely attributed to Richard Feynman, which states that if you think you understand it, then you don't understand it. If you follow the logic of this claim to its inexorable, bitter conclusion, it turns out that nobody understands the strange world of the universe's smallest entities. You should bear this in mind as you work through the puzzles featured here.

IS THE CAT DEAD OR ALIVE?

The life of Purccini, the famous stage cat, is under threat after an unfortunate accident that occurred as he was waiting to go on stage at Eigen Middle School. Mindful of the benefits of a Mediterranean diet, he had been tracking a peculiarly enticing fish smell when he found his way into a physics classroom. Here a large box grabbed his attention, and he threw himself inside it, causing its lid to slam shut.

Unbeknownst to Purccini, this is no ordinary box. It contains a Geiger counter, a tiny piece of radioactive material, a hammer and a sealed glass container full of cyanide. These are set up as follows. Over the course of an hour, there is an equal chance that one or zero atoms of the radioactive material will decay. If an atom does decay, then the Geiger counter will trigger a mechanism causing the hammer to smash the container holding the cyanide, thereby killing Purccini. If an atom doesn't decay, then Purccini will live on.

There are two other characteristics of the box worth mentioning. The first is that its design means that Purccini can neither interact with the mechanism nor escape, so he has no way of saving his own life. The second is that it's not possible to see inside the box, which means an observer cannot know whether an atom has decayed, and therefore cannot know whether Purccini is still alive. Exactly one hour after Purccini jumped into mortal danger, his owner, concerned that he has missed his curtain call, follows his muddy paw prints and tracks him down to the box.

Is Purccini dead or alive when his owner finds the box?

(SOLUTION ON PAGE 110)

HOW OLD IS EINSTEIN'S TWIN?

It's 1906, and Egbert Einstein, Albert's little-known twin brother, is in trouble. He is currently strapped into a spaceship, hurtling through deep space at more than three-quarters the speed of light, with only Purccini, the publicity- and fish-hungry stage cat, for company.

Egbert's trouble started when he gave an impromptu lecture at the *Gott und Dice*, a local Swiss hostelry. His subject was time dilation, and he'd had fun explaining how his brother Albert's new theory of special relativity showed that space and time are integrally connected; time is stretched depending on how fast an observer is traveling — put simply, the faster somebody goes, the slower time runs.

Egbert had gone on to point out that this has a number of bizarre consequences, perhaps the most interesting of which is that it allows for people to age at radically different rates. He illustrated this point with a simple thought experiment.

☞ Albert and Egbert, identical twins, are standing on the shore of Lake Geneva when a spaceship arrives, collects Egbert, and embarks on a round trip to the nearest star system, which is a distance of four light years away. The ship is traveling at 80% of the speed of light, which means in Earth time it will be ten years before Egbert is returned home.

However, at this sort of speed, significant time dilation will occur (time will run slower), which means only six years will elapse for those onboard the ship before they arrive back at Earth, a fact that will be confirmed by the ship's clocks and by the amount of biological aging that will have occurred.

This means that on his return, Egbert will be four years younger than Albert, despite the fact that they had been the same age when he was whisked away. 👉

Egbert had been successfully batting away objections until the intervention of Polizeikommissar Pferd, a regular visitor to the *Gott und Dice*. Pferd had pointed out that relativity entails that either of the twins could regard the other twin as the traveler. It seems natural to say that Egbert has whizzed off into deep space while Albert remains stationery. But, according to Einstein's theory, there is no frame of reference that defines things in this absolute sense, so Egbert can claim that it is Albert, the Earth and the distant star system that are moving relative to his spaceship. If this is the case then Albert, rather than Egbert, will be subject to time dilation, and when the twins are reunited, it'll be Albert who will be the younger of the two. This is a paradoxical result, because it can't be that both twins are younger than each other.

Egbert had been able to see the force of Pferd's argument, but annoyingly hadn't been able to refute it. It was at this point he had rashly suggested that they could test what would actually happen, if only they could find a twin daft enough to agree to be blasted into space at close to the speed of light…

Who will be the older of the two twins when Egbert returns to Earth?

(SOLUTION ON PAGE 112)

ARE THERE SPOOKY ACTIONS AT A DISTANCE?

It's 1935, and Albert Einstein has the feeling he's being haunted. Not by your average ectoplasmic entity, but rather by the ghostly realm of the universe's smallest particles. The problem is the newfangled quantum mechanics (hereafter QM), which seems to assert that the subatomic world is shadowy and nebulous until we observe it. In the absence of measurement, atoms, electrons, and the like have no determinate location, no specifiable momentum, but rather appear to be in many places and no place, all at the same time.

Einstein is not prepared to put up with this sort of wishy-washy thinking. After all, nobody really believes the moon is only there when somebody looks at it. So together with his colleagues Boris Podolsky and Nathan Rosen, he's come up with a thought experiment that attempts to show that QM cannot give a complete description of reality.

This is the essence of his thought experiment*.

Imagine that a particle splits into two fragments, and these fragments fly away from each other until they are galaxies apart. These new particles are entangled in such a way that their properties are inextricably linked to each other. If one of the particles is spinning in a clockwise direction, for example, it necessarily means the other particle is spinning in an counterclockwise direction. There's nothing unusual here; entanglement is a well-established and experimentally confirmed phenomenon.

The problem comes when entanglement is combined with the idea that until a definite observation or measurement occurs, the particles are in a state of superposition — that is, in terms of our spin example, both particles contain both clockwise and counterclockwise aspects. It is only when measurement is made that one of these aspects becomes real and the other drops away.

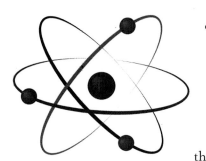

Let's say we measure the spin of particle A, and it turns out to be in a clockwise direction. So what happens to particle B, which is now many billions of miles away? Well, if it is entangled with particle A, such that its spin must be in the opposite direction, it must instantly bring its counterclockwise

* This treatment is based on Paul Davies's account in *God and the New Physics*.

aspect into being. To put this in technical terms, its wave function must collapse, revealing that its spin is anticlockwise. It has no other options.

But this all sounds entirely bizarre. How can B possibly know across the vast emptiness of space that A has realized its clockwise aspect? Moreover, even if information can be exchanged between the particles, how could that information travel faster than the speed of light (which instantaneity would require)?

Einstein calls this possibility "spooky action at a distance." The point of the thought experiment is to undermine the QM thesis that atoms, electrons, and the like have no determinate properties until they're observed. After all, the problem disappears if the particle fragments begin spinning in opposite directions at the moment of their creation.

Are there spooky actions at a distance, or do we have to give up on the QM idea that it's only measurement that brings the subatomic world properly into existence?

(SOLUTION ON PAGE 115)

WHEN IS A WAVE NOT A WAVE?

Thomas Young, scientist and polymath, is somewhat perturbed to find himself suddenly alive and well in the 21st century, and lying in what appears to be some kind of laboratory with a man in a white coat staring down at him. Last he could remember, it was the 19th-century, and he hadn't been feeling too well.

"Ah, Mr Young, you're back with us," says the white coat. "My name is Melquíades, I work for the Lazarus Corporation. You're probably wondering what you're doing here."

Melquíades tells Young that he's been brought back from the dead using the technique of retroactive cryogenics, quite normal in the late 21st-century, to witness an update to his famous double slit experiment.

Young has always been proud of this experiment. It was designed to establish whether light is corpuscular, comprising a flow of individual particles, or whether it takes the form of a wave.

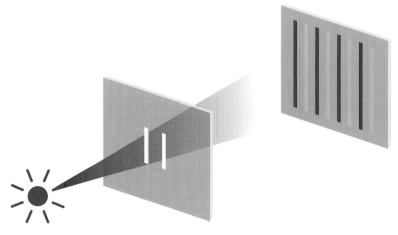

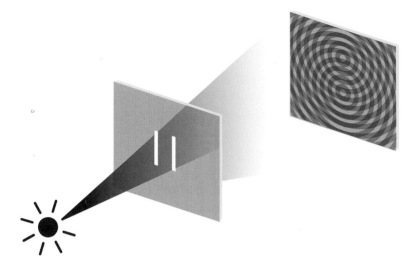

☞ If you shine a light at a screen with two slits (see above), and then capture the light on a plate placed behind the screen, you're going to get a pattern. If light is corpuscular, you'd expect two bands on the back screen that line up with the slits, as the particles will all travel in straight lines from the slits to the screen.

If light takes the form of a wave, then you'll see a pattern, a series of light and dark bands. This is because the waves will spread out after going through the slits, in a similar way to the circles of waves produced when you drop a pebble into a pond. As there are two slits there will be two waves, which will interfere with each other when they meet, causing a pattern of light stripes where peaks line up (constructive interference), and dark stripes where peaks and troughs cancel each other out (destructive interference). ☜

Young's experiment (see above) revealed an interference pattern, demonstrating that light has a wave-like character.

However, Melquíades explains to Young that since the early 20th-century, we have known that light is a little stranger than his experiment suggested, and is in fact made up of photons—tiny, indivisible packets of energy.

Young can't quite square this with the result of his experiment, but suggests to Melquíades that perhaps these photons obey a complex set of rules when acting in concert with each other as part of a flow of light, and that this explains the banding pattern he discovered. Not impossible, and not mysterious in any profound way.

Happily, in the 21st-century, it's possible to test this hypothesis by sending photons through the slits one by one, and this is precisely the updated experiment that Young has been resurrected to witness.

What will happen when we conduct this experiment? What should we expect to see?

(SOLUTION ON PAGE 118)

WHAT DID THE TELESCOPE REVEAL?

It's Saturday, March 5th 1870, and on the banks of the Old Bedford River the morning is bright and crisp, perfect conditions for John Hampden and Alfred Russel Wallace to settle a wager that is still being talked about to this day. At stake is £500.00, a small fortune in today's money, and the reputation of the British scientific establishment.

The events of this morning have their origins in an advertisement that Hampden placed in the journal *Scientific Opinion* in January of the same year.

> *The undersigned is willing to deposit £500, on reciprocal terms, and defies all the philosophers, divines and scientific professors in the United Kingdom to prove the rotundity and revolution of the world. He will acknowledge that he has forfeited his deposit, if his opponent can exhibit, to the satisfaction of any intelligent referee, a convex railway, river, canal or lake.* JOHN HAMPDEN

Alfred Russel Wallace, already well known for his independent discovery of the principles of natural selection, accepted the challenge, a decision he later called the biggest mistake of his life. He hoped to prove that the Earth is a sphere, a fact already accepted by nearly everybody in 19th-century Britain, by means of an experiment to be conducted between two bridges on a six mile stretch of the Old Bedford River. The experiment is as follows:

☞ A large cotton sheet with a thick black band painted along its center was fixed to the Old Bedford Bridge at a height of approximately thirteen feet above the water. Six miles away, a large

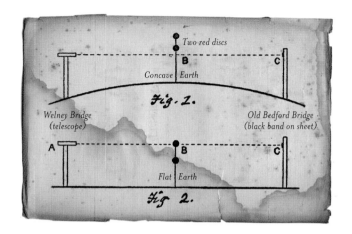

Two red discs

B   C

Concave | Earth

Fig. 1.

Welney Bridge   Old Bedford Bridge
(telescope)   (black band on sheet)

A   B   C

Flat | Earth

Fig. 2.

telescope was placed on the parapet of Welney bridge so that its line of sight was the same height above the water as the black band.

Halfway between the two bridges, three miles away from each, a pole with two red discs was fixed into the water, so that the top disc was the same height above the water as the black band (and telescope line of sight). The bottom disc was four feet lower down.

If the six miles of the river formed a convex shape — i.e., if the Earth is a globe — then the top disc would appear through the telescope to be substantially higher than the black band, as per the diagram (fig 1 above).

On the other hand, if the six miles of the river were flat — if the Earth is flat - then the top disc would line up with the black band on the Old Bedford Bridge (fig 2 above).

The question is straightforward. What did the telescope reveal when the test was conducted?

(SOLUTION ON PAGE 120)

4

PERPLEXINGLY PARADOXICAL

*...the way of paradoxes is the way of truth. To test reality we
must see it on the tightrope.*
OSCAR WILDE (*THE PICTURE OF DORIAN GRAY*)

The puzzles and conundrums featured here are perfectly designed to aid you in becoming the know-it-all in your own life. Bring them up at a dinner party and you'll be an instant star, or something like that.

Not all of them have solutions that are generally agreed upon. Those that do inevitably require some quick and tricky thinking. These puzzles are not for the faint of heart.

Let's see if you can help to thwart a desperate criminal's attempt to evade justice, sort out a supervillain's crisis of conscience, and figure out whether a pantheon of gods will prevent an arrogant and out of control tortoise from winning a race.

Although the puzzles are dressed up in shiny new clothes, most of them have a very long history. They remain nevertheless the focus of ongoing interest, which is both good news and bad news. Good, because if you manage to solve them, your name will be celebrated in the hallowed halls of academe. Bad, because you're probably not going to be able to solve them.

WHAT HAPPENS NEXT?

Dr Sandra Forster has had an eventful couple of years. She's moved out of her hauntingly beautiful Wisconsin lake house. Witnessed the death of a handsome Canadian stranger in a road traffic accident. Reconnected with an unsympathetic ex-boyfriend. And discovered a portal that allows her to send messages into the past. Not just any old portal, mind you, but her old mailbox. It turns out that the Wisconsin postal service is so efficient it's able to deliver a letter exactly two years before it was mailed.

Dr Forster, nobly eschewing the money-making potential of an ability to communicate with the past, has used the mailbox to cultivate a love affair with an architect who is living his life two years in her past and who has moved into the very same Wisconsin lake house. She sends him a letter, he sends her a letter, she sends him another letter; it's *84 Charing Cross Road*, with a bit of time travel added in.

Unfortunately, a large cloud has cast a malevolent shadow over the burgeoning love of our time-twisting paramours. Dr Forster had arranged a dinner date in her present with Mr Architect, but he hadn't turned up. She just can't understand it. If Penelope could wait 20 years for the return of Odysseus, and reject the advances of 108 different suitors, then surely Mr Architect should have been able to navigate the Scylla and Charybdis of two years without too much mishap.

But as it turns out, mishap doesn't cover even the half of it. Dr Forster has just today discovered the terrible truth. The reason Mr Architect didn't show at their dinner date,

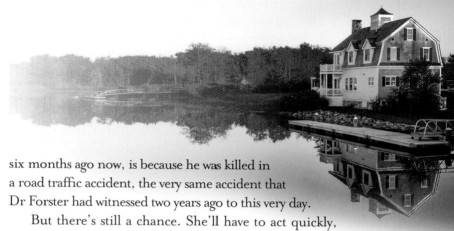

six months ago now, is because he was killed in
a road traffic accident, the very same accident that
Dr Forster had witnessed two years ago to this very day.

But there's still a chance. She'll have to act quickly,
but if she can get a letter to him at the lake house then maybe
she can prevent him from taking his final fateful trip. It seems
that he had figured out where she had been working two years
ago, and was on his way to see her. The cruel hounds of love!

So Dr Forster writes her letter. She explains that he
mustn't attempt to find her, that he'll be killed in a road
traffic accident if he does, and that she's here right now at
the lake house, and if he'll just wait two years, he can come
to find her, and they can be together forever. She places her
letter into the mailbox, and snaps its door shut. She has sent
her message into the past.

What happens after Dr Forster sends her letter into the past?

(SOLUTION ON PAGE 130)

5

LIFE, THE EARTH AND EVERYTHING

*The eternally incomprehensible thing about the world
is its comprehensibility*
ALBERT EINSTEIN

Congratulations! You've made it to the final chapter, and hopefully you're still in good shape. Are you ready for your final challenge? This last chapter shifts focus towards matters that are somewhat more concrete in form.

Ever wondered why an alien life form has never popped up on the evening news to announce its arrival on Earth? If so, you're not alone, as you'll discover when you wrestle with the Fermi paradox. Maybe you have heard it said that the humble bee defies the laws of physics when it takes to the sky? Perhaps that sounds a little implausible to you, but do you know exactly how the bee does fly?

Probably you'll be pleased to hear that only one of the puzzles in this chapter is genuinely intractable. It's a curious example of a time-travel paradox that doesn't result in a logical contradiction, but nevertheless produces results that are highly counterintuitive. Maybe you can figure it out, though. After all, if you've worked your way through all the puzzles in this book, you have by now had plenty of practice making sense of the apparently nonsensical.

WHERE ARE ALL THE ALIENS?

Badger Blight is a little ticked off. He's been hanging around Area 51 in the middle of the Nevadan desert for the last three years, and he's not seen a single alien. No little green men, no flying saucers, not even an alien autopsy. It's all been a massive disappointment. He really can't understand it. It is well known among Ufologists that all the best aliens hang out at Area 51, but now that he's turned up, they've apparently hightailed it back to their spacecraft and vamoosed.

To make matters worse, he's being teased mercilessly by his oh-so-skeptical friend, Danielle Scullery. She just laughed at him when he explained about the government conspiracy to keep alien visits a secret, and she's started calling him Mr Grey for reasons he suspects have something to do with the double star Zeta Reticuli. She also keeps banging on about what she calls the Fermi paradox, which she explains as follows.

The Earth is part of a relatively young planetary system within a galaxy called the Milky Way, which is 13.5 billion years old. Latest research suggests that the Milky Way is home to at least 100 billion planets, probably more. Even if intelligent life has evolved on only a tiny fraction of these planets, that's still an awful lot of intelligent life. An alien civilization that has mastered rocketry could quite easily colonize the Milky Way. It might take 100 million years, which sounds like a long time, but actually represents less than 100th of the age of the galaxy. If there are technologically advanced alien civilizations, it's hard to believe that not one of them would be interested in colonizing space.

Badger does not like this line of argument, not least because three years ago he would have believed the Earth had already been visited by extraterrestrials. But now he's not so sure. If they're not at Area 51, where are they exactly? Danielle Scullery, for her part, is fairly certain she knows how to resolve the Fermi paradox. No aliens have visited Earth, because there are no aliens. Human beings are alone in the galaxy.

So who is right? If the galaxy is teeming with intelligent life, why hasn't it turned up on Earth? Perhaps there is an explanation for the absence of aliens that does not rely on denying the possibility of intelligent extraterrestrials. After all, it is quite hard to believe that on a small blue planet in an unremarkable corner of the Milky Way, intelligent life evolved for the first, and so far the last, time.

But that is precisely what makes this all a bit puzzling.

If the Earth is not alone in the galaxy in sustaining intelligent life, where are all the aliens?

(SOLUTION ON PAGE 140)

SOLUTIONS

MARRIED OR NOT?

(see pages 10–11)

The registry office logic test set by Inspector Horse is a version of a puzzle originally devised by Hector Levesque. It seems easy to get right. Most people answer that we cannot determine whether a married person is looking at an unmarried person.

Unfortunately, most people answer incorrectly. The right answer is "Yes" — a married person is looking at an unmarried person. Let's examine the question in detail so we can see why "Yes" is the right answer.

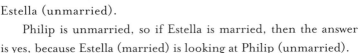

The set-up is straightforward:
Bentley is looking at Estella, but Estella is looking at Philip. Bentley is married, but Philip is not.

And so is the question:
Is a married person looking at an unmarried person?

Bentley is married, so if Estella is unmarried, then the answer is yes, because Bentley (married) is looking at Estella (unmarried).

Philip is unmarried, so if Estella is married, then the answer is yes, because Estella (married) is looking at Philip (unmarried).

We don't know whether Estella is married or not, but it doesn't matter — either way, a married person is looking at an unmarried person. If she's married, then she's looking at an unmarried person (Philip). If she's not married, then a married person (Bentley) is looking at her.

Therefore, the correct answer is yes, a married person is looking at an unmarried person.

Why do we get it wrong?

Keith Stanovich, a psychology professor at the University of Toronto, estimates that some 80% of people get this question wrong. The reason why most likely has to do with the fact that human beings tend to be cognitive misers. Put simply, we tend to take intellectual shortcuts if they're available and if we're not on our guard.

In this case, we probably reason as follows: Bentley is married, but we don't know whether Estella is married or unmarried, so we can't determine what's happening here. Philip is unmarried, but again we don't know Estella's marital status, so we can't determine what's happening here either.

It's at this point our cognitive miserliness kicks in, and we make the logical error. We assume that because we don't know in either of these cases whether a married person is looking at an unmarried person, it follows that we cannot know that a married person is looking at an unmarried person. It doesn't follow, of course, because if it isn't true in the first of these cases, it must be true in the second (and vice versa). But we never get this far; we stop reasoning once we see that we do not know whether Estella is married, and leap straight to the easiest conclusion.

Don't feel too bad if you got this question wrong. In a sense, we're supposed to take shortcuts in reasoning. Shortcuts are useful because they allow us to make effective use of our limited cognitive resources, and most of the time they work. It's just that occasionally, perhaps more often than we'd like, they lead us astray.

ARE YOU THE NEW SHERLOCK HOLMES?

(see pages 12–13)

Jack Dawe's test is a measure of your ability to assess the validity of a syllogism — an argument with two premises and a conclusion. Here, "validity" means something very precise: an argument is valid if its conclusion is logically required by its premises.

This is the standard example of a valid argument:

All men are mortal

Socrates is a man

Therefore, Socrates is mortal

If all men are mortal, and Socrates is a man, it follows as a matter of logical necessity that Socrates is mortal.

Perhaps the most important thing to realize about the idea of validity is that it's possible for a valid argument to have a false conclusion. For example, consider the following argument:

All men are immortal

Socrates is a man

Therefore, Socrates is immortal

This is a valid argument (the conclusion is logically entailed by its premises), but the conclusion is false (because the first premise — All men are immortal — is false).

Let's have a look at the arguments that you had to assess.

ARGUMENT I

Premise: All things that are swallowed are good for the health

Premise: Alcohol is swallowed

Conclusion: Alcohol is good for the health

This argument is valid. The conclusion is dubious at best, but it follows as a matter of necessity from the premises, which means it's a valid argument.

SOLUTIONS

ARGUMENT 2

Premise: Every animal with a tail is vicious

Premise: A golden retriever is not vicious

Conclusion: A golden retriever is not an animal with a tail

This argument is also valid. If it were true that all animals with tails are vicious, and also that a golden retriever is not vicious, then it necessarily follows that a golden retriever is not an animal with a tail.

If you don't understand why this is the case, try the same argument with different terms. For example:

Premise: All humans have brains

Premise: A table doesn't have a brain

Conclusion: A table isn't human

ARGUMENT 3

Premise: All short people have orange skin

Premise: Ronald Plump is not short

Conclusion: Ronald Plump does not have orange skin

This argument is not valid. The conclusion may or may not be true, but either way it doesn't follow from the premises.

Belief Bias

All of the arguments featured here have been designed to illustrate something called belief bias. Psychologists such as Jonathan Evans have demonstrated that we're not great at working out whether deductive arguments are valid when the final conclusion runs contrary to our beliefs. Put simply, we tend to get distracted by the truth (or not) of the conclusion, and don't pay sufficient attention to the logic of the argument. So if you've done badly, you could always try telling Jack Dawe that it's not that you're bad at logic, it's just that you fell victim to a common cognitive bias.

SWITCH OR STICK?

(see pages 14–15)

Jack Dawe is wrestling with a version of the Monty Hall puzzle, a notoriously fiendish test of logic. Nearly everybody who sees this puzzle for the first time supposes there is no advantage to switching from the original choice. It seems obvious, given our scenario, that no button is more likely than any other button to be the escape button, which means there is nothing to be gained by switching. Once the first trigger button has been pushed, it's fifty-fifty between the remaining two buttons as to which is the escape button.

Although this line of argument seems perfectly reasonable, it's actually wrong. Jack Dawe should switch from his original choice if he wants to maximize Lola's chance of survival. Here's why.

The best way to approach this puzzle is to recognize that it's impossible for Jack Dawe to lose by switching (i.e., to pick a trigger button) if his original choice is itself a trigger button. If Dawe starts off by selecting a trigger button, then the gang member is forced to push the other trigger button, which has the effect of revealing where the escape button is — it's the button he doesn't push. The probability that Dawe will start off by choosing a trigger button is 2-in-3, which means that by switching he has a 2-in-3 chance of winning (and it follows only a 1-in-3 chance of winning if he doesn't switch).

Let's run through the various possibilities to make this absolutely clear.

POSSIBILITY 1 (LOSE)

Dawe starts off by selecting the escape button. Gang member pushes one of two trigger buttons. Dawe switches to the other trigger button. Lola is blasted into the stratosphere.

POSSIBILITY 2 (WIN)

Dawe starts off by selecting trigger button 1. Gang member pushes trigger button 2. Dawe switches to the escape button. Lola makes a run for it, and lives.

POSSIBILITY 3 (WIN)

Dawe starts off by selecting trigger button 2. Gang member pushes trigger button 1. Dawe switches to the escape button. Lola hoofs it out of danger, and lives.

Switching results in two wins and one loss, which means that not switching results in two losses and one win.

Jack Dawe should switch if he wants to give Lola her best chance of avoiding being transformed into aerial mutton.

HOW WIDE IS THE LAKE?

(see pages 16–17)

Captain Trink is wrestling with a version of a problem first posed by American puzzler Sam Loyd more than one hundred years ago. Loyd said of his puzzle that though ninety-nine out of a hundred of the shrewdest business men would not figure it out in a week, in fact it only requires common sense and simple addition to solve. This is how you get to the right answer.

Combined Distances

The key to solving the puzzle is to focus on the combined distances traveled by both ships, and then to use this data to work out how wide the lake is — and how far the *Tribulation* has traveled.

At the point at which the ships first meet, they have traveled a combined distance that is equal to the width of the lake (because they started on opposite banks). We also know the *Tribulation* has traveled 720 yards.

At the point at which they next meet, the ships will have traveled a combined distance that is equal to three times the width of the lake. It is not too difficult to understand why. Both ships have completed one width of the lake (which equals a combined two widths), and then they meet again on their return journey (so you add another width to the two you've already got).

It follows that at their second meeting the *Tribulation* has traveled three times as far as it had at the first meeting. This is true, because it is specified that both the *Tribulation* and *Rapide* are traveling at a constant speed, and three combined widths is three times as far as the one combined width the ships had traveled when they first met.

We now have all we need to calculate the width of the lake. We know that the *Tribulation* has travelled 2160 yards by the time it meets the *Rapide* on its return journey (3 x 720 yards). We also know that it has traveled 400 yards more than the length of one width of the lake (because it's 400 yards away from the shore from which it set off on its return journey). So if we subtract 400 yards from 2160 yards we get the width of the lake.

2160 - 400 yards = 1760 yards, which is exactly one mile. This means that Lake Chudleigh is one mile wide, and that the *Tribulation* will have covered two miles by the time it gets back to its original starting point. On the basis of this data, Captain Trink would have been able to calculate the average speed of his ship, if only he hadn't forgotten to time the journey. He fears there might be a rendezvous with a plank in his not too distant future.

HOW MANY COOKIES ARE LEFT?

(see pages 18–19)

The puzzle involving Reba and the cookies, and the question of how many will be left after she has completed her journey to John O'Groats, is a version of a mathematical brain teaser that normally features a camel and 3000 bananas.

The puzzle itself is simple. Reba needs to transport 3000 cookies a distance of 1000 kilometers. She will eat one cookie for every kilometer she travels, and the maximum number of cookies she can carry at the same time is 1000. What is the largest number of cookies she can get to her final destination?

It should be immediately obvious that Reba can't just pick up 1000 cookies, and head for John O'Groats. She'll eat the final cookie as she arrives at her destination. The optimal strategy must involve a staggered approach. Put simply, Reba needs to go back for the cookies that she can't carry in a single journey, ensuring that there are cookies available for her return journeys.

Staggered Journey

Let's see how this might work traveling one kilometer at a time. Reba picks up 1000 cookies, moves them one kilometer, eats a cookie, and then goes back to pick up a second lot of 1000 cookies, eating another cookie as she arrives back to pick them up (because she's travelled another kilometer).

> *Total Cookies Consumed: 2*
>
> *Total Cookies Transported: 1000*

Reba picks up the second lot of 1000 cookies, and exactly repeats this process, arriving back to pick up the final batch of 1000 cookies.

Total Cookies Consumed: 4

Total Cookies Transported: 2000

She now picks up the final batch, and takes them one kilometer, where they are reunited with all the other cookies. She eats another cookie, but obviously doesn't need to go back this time, so that's it for this leg of the journey.

Total Cookies Consumed: 5

Total Cookies Transported: 3000 (of which 2995 remain)

Total Distance Travelled: 1 kilometer

As you can see, when Reba has to make three journeys to transport all the cookies, it's going to cost a total of five cookies for each kilometer traveled (which is why she only has 2995 cookies left after one kilometer).

She has to make three journeys for as long as there are more than 2000 cookies remaining. At a rate of five cookies per kilometer, she'll be down to 2000 cookies once she hits the 200 kilometer mark (3000 cookies minus [5 200 cookies] = 2000 cookies).

At this point, each kilometer is going to cost less in cookies, because Reba only needs to make two journeys per kilometer. Let's see how this works.

Reba picks up 1000 cookies, moves them one kilometer, eats a cookie, and then goes back to pick up the second, and final, batch of 1000 cookies, eating another cookie as she arrives back to pick them up. She picks up the second batch of cookies, and takes them one kilometer, eating another cookie as she arrives to reunite them with the first lot of 1000 cookies. She has now transported all the remaining cookies another kilometer, at the cost of three cookies.

As you can see, when Reba has to make two journeys to transport the cookies, it's going to cost a total of three cookies for each kilometer traveled.

She has to make two journeys for as long as there are more than 1000 cookies. (Obviously, once she hits the 1000 cookie mark, she can just pick up the whole lot, and go on her merry way.)

At a rate of three cookies for each kilometer, she'll be down to 1001 cookies after 333 kilometers (2000 cookies minus [3 x 333 cookies] = 1001 cookies), at which point Reba should treat herself to an extra cookie (because it doesn't

make sense to come back for this one additional cookie), leaving her with 1000 cookies.

She has now traveled 533 kilometers (200 + 333), leaving her with 467 kilometers to go, and 1000 cookies to fuel her journey.

She should now just pick up the remaining 1000 cookies and complete the journey, eating one cookie for every kilometer she travels. This means she'll have 533 cookies (1000 - 467) left to hand over to the baker in John O'Groats. Hopefully, this will be enough cookies to secure her future at Whippet Over, Bovey's most successful delivery company staffed only by dogs. If not, though undoubtedly an honourable hound, she suspects the remaining 533 cookies might mysteriously disappear.

WHO LOST WHAT?

(see pages 20–21)

This puzzle was originally posed by Lewis Carroll, best known as the author of *Alice's Adventures in Wonderland*. Carroll was a mathematician by profession, and a keen logician and puzzler. The puzzle appears in Edward Wakeling's 1995 collection of rediscovered Lewis Carroll puzzles.

There is an obvious solution to the puzzle that is not too difficult to work out. Let's track the various transactions to see what's happened.

Gibbon starts with I shilling, which is 12 pence. He gives this to the pawnbroker, and gets 9 pence back and also a pawn ticket for the shilling.

Gibbon	Pawnbroker	Friend
9 pence	I shilling	n/a
Pawn ticket	Minus 9 pence	

At this point, Gibbon has 3 pence less than he started with, but he has the pawn ticket. The pawnbroker has paid out 9 pence, but he's holding an item of greater value (i.e., the shilling).

Gibbon then sells the pawn ticket to a friend for 9 pence.

Gibbon	Pawnbroker	Friend
9 pence	1 shilling	Minus 9 pence
9 pence	Minus 9 pence	Pawn ticket

As you can see, we're now in a situation where Gibbon has 18 pence (6 pence more than he started with), and his friend is 9 pence down, but the friend now owns the pawn ticket.

Imagine now the friend goes off to the pawn shop to get the shilling back. Obviously, the pawnbroker isn't just going to give it to him. The friend will have to pay for it. Thus, we get:

Gibbon	Pawnbroker	Friend
9 pence	Minus 9 pence	Minus 9 pence
9 pence	Plus 9 pence	*(to Gibbon)*
	(from friend)	Minus 9 pence
		(to Pawnbroker)
		Plus 1 shilling
		(from Pawnbroker)

If we do the math, we'll see that Gibbon has 18 pence (6 pence more than he started with), and his friend has 1 shilling (6 pence less than he has spent across the two transactions).

The answer is that the friend has in effect given Gibbon 6 pence, which explains how Gibbon is now able to go to see the Bovey Amateur Dramatics Society's production of *Les Misérables*.

Of course, in the real world it wouldn't be quite so straightforward. In this set of transactions the pawnbroker has ended up exactly where he started – he gave 9 pence to Gibbon, and then received 9 pence from Gibbon's friend. But pawnbrokers do not offer their services for free. In this transaction, the pawnbroker would have charged some kind of fee, which would be specified on the pawn ticket. Therefore, though we cannot know exactly how much money Gibbon's friend has lost in the transaction, we do know that it's more than 6 pence (and no more than 9 pence – obviously Gibbon's friend is not going to pay more than a shilling to get the shilling back from the pawnbroker).

The moral of the story: don't do business with itinerant ukulele virtuosos when there's a performance of "I Dreamed a Dream" in the reckoning.

HOW OLD IS HE?

(see pages 22-23)

The Reverend Daniel Martin has been set what is known as an Age Problem. It is possible to adopt a "brute force" approach to solve these problems, which involves guessing ages until you hit upon a combination that works for the problem under consideration. Trouble is, if you're not running through the combinations using a computer, you'll likely be around for a while before you hit upon the right answer.

By far the best way to solve these kinds of puzzles is to use tables to conceptualize the problem, and algebra to solve it. Let's see how we'd use this approach to calculate Charon's age.

First off, let's set up a table.

	Age now	Age in 2 years
Charon		
William		

Now let's start to fill it in with the information we know.

	Age now	Age in 2 years
Charon	x	$x + 2$
William	$32 - x$	$(32 - x + 2) = 34 - x$

We don't know Charon's age, so we let it be x. We don't know William's age either, but we do know that it'll be 32 - x (because we're told that the sum of the ages of Charon and William is 32). This means in the Age Now column we want x for Charon and 32 - x for William.

The Age in 2 Years column is straightforward. We've got to add 2 years to both ages (which, as you can see above, we've done).

Algebraic Manipulation

Now we've got to do a bit of basic algebra. Don't worry, this is pretty easy. Let's go back to our table, and substitute 25 for x in the Age Now column.

Charon = 3* William	In two years, Charon is three times as old as William.
$(x + 2) = 3(34 - x)$	This is just Charon's age in 2 years, which equals 3 times William's age in 2 years.
$x + 2 = 102 - 3x$	Get rid of the parentheses. (Basically, we've just multiplied 34 and x on the right hand side by 3).
$4x + 2 = 102$	We can get rid of the 3x on the right by adding 3x to both sides (so we end up with 4x on the left).
$4x = 100$	Dead easy. Just take 2 from both sides.
x = 25	If $4x = 100$, then $x = 25$.

	Age now	Age in 2 years
Charon	x	$x + 2$
William	$32 - 25 = 7$	$(32 - x + 2) = 34 - x$

And there's your answer. The fishy fellow Charon is 25 years old.

WHO WILL BE EXECUTED, WHO WILL BE PARDONED?

(see pages 26-27)

Grace, Chilly Creek Farm's tastiest turkey, is wrestling with a version of a puzzle known as the Three Prisoners Problem, which was originally posed by mathematician Martin Gardner in 1959.

Gardner called the puzzle "bewildering," and the logic that leads to the correct answer is indeed convoluted and discombobulating, but we'll do our best to make things clear.

The best approach is to work out the probability of each of the possible responses that Reuben can give to Grace. To recap, this is the formula governing Reuben's response:

Pardoned	To be executed *
Grace	Amy or Freckles (random choice)
Amy	Freckles
Freckles	Amy

* according to Reuben

We already know that the prior probability of any particular turkey being pardoned is 1-in-3. It was a random selection. This allows us to work out the probability of each of Reuben's four possible responses.

The Amy and Freckles cases are straightforward enough. There's a 1-in-3 chance that Amy will be pardoned, which means there's a 1-in-3 chance Reuben will say Freckles is to be executed *because Amy is to be pardoned* (remember his choice is forced here, he can't say Grace). And it's the other way around if Freckles is to be pardoned; that is, there's a 1-in-3 chance that Reuben will say Amy is to be executed because *Freckles is to be pardoned.*

It's a little bit more complicated if Grace is to be pardoned.

The probability of Grace being pardoned is also 1 in 3. If Grace is to be pardoned, then half the time Reuben will say Amy is to be executed, and half the time he'll say Freckles is to be executed. This means the probability that Reubens will say Amy is to be executed *because Grace is to be pardoned* is 1 in 6 (Đ ½), and likewise the probability he will say Freckles is to be executed *because Grace is to be pardoned* is also 1 in 6.

Pardoned	To be executed *	Probablity
Grace	Amy	1/3 x 1/2 = 1/6
Grace	Freckles	1/3 x 1/2 = 1/6
Amy	Freckles	1/3
Freckles	Amy	1/3

according to Reuben

We can now solve the puzzle of whether Grace is right to think it's fifty-fifty between herself and Freckles for the chop. Remember, Reuben has told Grace that Amy is to be executed. According to the table above, the probability of Reuben saying that Amy is to be executed because Freckles is to be pardoned is 1 in 3. Unfortunately for Grace, the probability of him saying Amy is to be executed because Grace is to be pardoned is only 1 in 6.

This means the response that Reuben gave is twice as likely to be elicited because Freckles is to be pardoned than it is because Grace is to be pardoned. To put it another way, given Reuben's response, Freckles will be pardoned 2 out of 3 times, whereas Grace will be pardoned only 1 out of 3 times.

The chance that Grace will be executed is not fifty percent. It's 1 in 3, exactly as it was when we started.

HOW LONG WILL THEIR RELATIONSHIP LAST?

(see pages 28-29)

Apollo has taken great delight in letting courting couples know when their relationships are likely to break down. The important question, of course, is whether his predictions are accurate. Does he get it right?

The surprising, though gratifying, answer is that in certain circumstances the method he uses is likely to generate accurate predictions. It's based on a paper by J. Richard Gott III that appeared in the journal *Nature* in 1993.

Gott hit upon his method during a visit to the Berlin Wall in 1969, eight years after the wall had first been erected. He began to wonder how long the wall would last. Using the same reasoning that Apollo employs he calculated that there was a 50% chance the wall would last more than 2 2/3 years but less than 24 years. Sure enough, the Berlin Wall came down in 1989, 20 years after he made his prediction, and within the time frame he predicted.

Gott also used his method to predict the future of 44 Broadway and Off-Broadway productions, using a 95% confidence limit. At last count, 36 of the 44 had shut down, and all had shut down in line with his predictions.

So the method works, but there are caveats — and they are important to understand. First, the thing about which you're making a prediction must be of indeterminate longevity. For example, you could use Gott's method to predict how long a magazine you've just come across will last given the date of its first issue, but you can't use it to generate an accurate prediction of the longevity of a fixed-term presidency (not that you'd need to).

Second, you have to encounter the thing at a random point in its duration. You can't turn up at a wedding hoping to predict how long the marriage will last, because you're at an event precisely to celebrate its beginning. To put this another way, you cannot enjoy any sort of special relationship to the history of the thing under consideration. If your marriage is in trouble, you can't use this method to work out how long it will last, because your enquiry is not indifferent or random — it's motivated by something that is directly implicated in its longevity. Similarly, it would have been no good rushing to the Berlin Wall as it was being torn down just so you could reassure the Soviet authorities that it was going to stand for another five years at least.

It's also worth noting that we've generated predictions that should be right 50% of the time. But there's no reason why you can't choose other levels of confidence. The math is different, but the same underlying principles are in play. As mentioned before, Gott uses 95% confidence limits for some of his predictions. This means that the ranges are larger, but you can be more confident they'll turn out to be accurate.

Let's finish by doing a calculation. It's 244 years since the United States declared independence. Using a 95% confidence level, we can predict that the United States of America will last for at least another 6 1/4 years, but for fewer than 9516 years.

RESPONSES

ARE WE LIVING IN A SIMULATION?

(see pages 30-31)

Nick Bostrom's simulation argument rose to prominence around the turn of the new millennium. It's important to get clear exactly what it claims to show. The argument is not an attempt to show that we're living in a simulation. Rather, it aims to demonstrate that we must accept, on pain of inconsistency, the truth of at least one of the three following propositions.

(1) It is overwhelmingly likely that intelligent species will go extinct before they are technologically mature.

(2) Virtually no technologically mature civilizations will be interested in running computer simulations of intelligent minds.

(3) If you're an intelligent being, you're almost certainly living in a simulation.

According to Bostrom, although at least one of these propositions must be true, it's not certain which that is. Take the first proposition, for example. It's not difficult to imagine that intelligent species might destroy themselves before they become technologically mature. Humanity's current dalliance with weapons of mass destruction is instructive in this regard.

But what about the argument itself? Are we compelled to accept the truth of at least one of these propositions? That isn't at all clear. There are several ways this argument can be challenged.

Simulation Objection: *If we are living in a simulation, then Bostrom's argument cannot get started. We don't know anything at all about what's happening outside of our simulation, so we can have no opinion about propositions (1) and (2).*

Bostrom's response to this objection is to say that if we're living in a simulation, then (3) is true. Well yes, clearly, but not for any reason that has to do with the argument.

Chinese Room Objection: *The simulation argument rests on the possibility of simulating a mind within a computer. Not everybody accepts that this will ever be possible. The philosopher John Searle, for example, developed a thought experiment called The Chinese Room, which purported to show that a digital computer, executing a program, can never manifest consciousness or understanding.*

If Searle is right, then there is zero chance we're living in a digital computer simulation. Arguably this would mean that Bostrom's proposition (2) is true — that is, no technologically mature civilizations will be interested in running simulations of minds — but that hinges on what exactly one takes "interested" to mean.

Infinite Universe Objection: *If the universe is infinite, then it might contain infinite simulated and infinite non-simulated minds. This messes with the math, because the ratio of simulated to non-simulated minds cannot be specified, so it's not possible to conclude that it's much more likely we are simulated rather than non-simulated.*

Bostrom is aware of this objection, and suggests there are a number of ways that one might deal with the difficulty.

Conclusion

Even if it turns out that Curtis exists only in a simulation, in terms of his day-to-day life it doesn't make much difference. Curtis's reality is a simulated universe. All the ordinary methods for organizing his life will still work. Assuming no intervention from the simulators, his simulated world will continue to function according to the laws that govern it. This is not to say that the knowledge he has of his true situation will make no difference. It will certainly provide a different perspective on his place in the world, and his world's place in a universe that includes at least one technologically mature civilization capable of simulating minds.

WHICH DIE SHOULD SHE CHOOSE?

(see pages 32-33)

Grace, that most charming of turkeys, is participating in a desperate game of chance with her very life at stake. She has to choose one die out of three to maximize her chances of winning a best-of-ten-throws contest.

These are the dice.

Die A has sides 2, 2, 4, 4, 9, 9

Die B has sides 1, 1, 6, 6, 8, 8

Die C has sides 3, 3, 5, 5, 7, 7

So which should she choose?

The answer is that it makes no difference. Regardless of her choice, the evil Dr Stuffing can pick a die that will win more than half the time. This is because he is using a set of what are called non-transitive dice for the contest. Put simply, Die A will beat Die B more than half the time. B will beat C more than half the time. And C will beat A more than half the time, despite the fact A tends to beat B, the die that tends to beat C.

This means the person who chooses second can always pick a die that will win on average. If Grace picks A, Dr Stuffing will pick C. If she picks B, he'll pick A. If she picks C, he'll pick B.

Here are the possible outcomes for these scenarios (opposite page). As you can see, it doesn't matter which die Grace chooses. Dr Stuffing always has the superior win ratio. The only way Grace can maximise her chance of winning is if Dr Stuffing chooses his die first. Otherwise, he has the upper hand. The dice are fair and balanced, but the game is rigged.

	Dr Stuffing (c)		
	3	5	7
Grace (A) 2	S	S	S
4	G	S	S
9	G	G	G

Dr Stuffing win ratio is 5/9

	Dr Stuffing (A)		
	2	4	9
Grace (B) 1	S	S	S
6	G	G	S
8	G	G	S

Dr Stuffing win ratio is 5/9

	Dr Stuffing (B)		
	1	6	8
Grace (C) 3	G	S	S
5	G	S	S
7	G	G	S

Dr Stuffing win ratio is 5/9

However, there is a glimmer of hope for Grace. If the contest were to be decided on the basis of one throw of the dice, she'd have a fairly close to evens chance of winning (44%). Even in a contest across 10 throws, Grace has a 35% chance of winning. Not great odds, but not yet a death sentence.

While the gods do like to kill us for their sport, here's to hoping they are smiling upon Grace this Thanksgiving as she attempts to vanquish the evil Dr Stuffing.

HOW MANY POSSIBLE WORLDS?

(see pages 34-35)

Marlon Malloy is not the first person to respond with incredulity on being asked to accept the existence of other possible worlds that contain counterpart versions of ourselves. On the face of it, the idea sounds far-fetched, but in fact it forms part of a respectable, though minority, metaphysical position called modal realism, developed by American philosopher David Lewis.

The big reason philosophers invoke possible worlds is that it allows us to talk sensibly about possibility, necessity, contingency, and so on. That sounds a bit highfalutin, but actually isn't too complicated. The idea that Hubert Humphrey could have won the 1968 presidential election (which he lost to Richard Nixon), for example, can be understood as amounting to the claim that there exists at least one possible world in which Humphrey won the election. Similarly, to say that the proposition "all unmarried men are bachelors" is necessarily true amounts to saying that it is true in every possible world.

So far so good. None of that sounds particularly unreasonable. But things get a bit weird when David Lewis steps onto the stage. Most people will probably suppose that these possible worlds are imaginative constructions. Useful, but not actually real. But Lewis insists they exist in just the same way as our actual world exists. They might look radically different to our own world, they might differ in content, but they are the same kind of thing.

So what's his justification for believing that possible worlds are as real as our actual world? Well, it's partly to do with the modal statements we discussed above. Lewis thinks that possible worlds can only help us to make sense of concepts such as possibility and necessity if they are real. As purely imaginative constructs they don't do the job. He also notes that we have no problem in asserting the existence of theoretically useful objects in the domain of mathematics (such as sets), so we should have no problem in doing the same in the case of possible worlds.

Likely this is not going to persuade you, just like it hasn't persuaded many professional philosophers. But modal realism has proved quite resilient, and it is hard to falsify. Nevertheless, there are things to be said against it.

Perhaps the most significant objection is that the theory is almost drunkenly inflationary about worlds. Modal realism holds that there are an infinite number of possible worlds in which additional possibilities can be realized. As Lewis recognized, this is likely to provoke incredulity, and at the very least the modal realist has to do some fast talking to persuade a critic that the theory doesn't fall foul of the principle that simpler explanations are better explanations.

Where does this leave us? Modal realism remains a live issue within professional philosophical circles. Therefore, it's a bit much to hope that Marlon will conjure up a knockdown argument to bring an end to the modal madness. But perhaps he could point out that it really isn't much consolation to him that his counterpart became a contender. He is not his counterpart, so he doesn't get to experience that glorious possible world.

WHAT'S THE REAL BENEFIT?

(see pages 36-37)

Basil Sinclair is about to start a course of lavoisian pills to help combat phlogiston poisoning. He has been persuaded to do so by some clinical studies that show that lavoisian reduces the risk of heart attack caused by a buildup of phlogiston by between 36% and 54%. Does this evidence provide sufficient grounds for believing that the therapeutic effect of lavoisian is significant?

Suppose you learn that a study has been conducted which shows that wearing a giant rubber hat while outdoors reduces the risk of being killed by a piece of falling space debris by 36%? Would you be persuaded that this is a good reason to wear such a hat?

Most people will say no, for the simple reason that the chance of being hit by a piece of space debris in the first place is miniscule. It might be true that you're less likely to be killed if you're wearing the hat relative to the risk of being killed if you're not wearing the hat, but because there is virtually no chance of being killed in either case, the improvement in your odds doesn't amount to much.

The comparison between the two groups tells you something about the relative risk of wearing a rubber hat compared to wearing no hat. But it tells you nothing about absolute risk. If you don't know the absolute risk, then you don't know whether the reduction in relative risk is something to which it is worth attending.

Let's apply this insight to Basil's worrying case of phlogiston poisoning. We know that taking lavoisian results in a relative risk reduction of between 36% and 54%. But we have no idea what this means in absolute terms. We don't know how often a buildup of phlogiston leads to a heart attack, so we don't know how many people would have to take lavoisian before one of them avoids a heart attack that they would otherwise have had. To put it simply,

if only one in five hundred thousand people poisoned with phlogiston suffers the complication of a heart attack, you're going to have to administer lavoisian to a truckload of phlogiston cases before you save one life.

A better measure than relative risk reduction to measure the efficacy of a drug intervention is a statistic known as "number needed to treat" (NNT). This refers to the number of people you need to treat over a given time period for one person to benefit. In our example, a person benefits if they don't have a heart attack that they would otherwise have had. Basil doesn't know the NNT for lavoisian, so he does not have enough information to make a judgment about whether its therapeutic effect is significant.

At this point we should note that phlogiston poisoning isn't a real thing and lavoisian doesn't exist. However, the studies cited above, showing a relative risk reduction of between 36% and 54%, are real. They provide part of the justification for the widespread use of a very well-known medication. It might surprise you to learn that in the case of the first study, which was conducted over a period of more than three years, the NNT for one person to benefit is 100. To put it simply, to avoid one bad outcome, 100 people need to take the drug for at least three years. The 36% relative risk reduction seems impressive, but the devil is in the detail. If you don't know what's going on in absolute terms, then you don't have enough information to make a judgment about whether a particular therapeutic effect is significant.

WHOSE BIRTHDAY IS IT ANYWAY?

(see pages 38-39)

Clare's troubles with the Archbishop of Canterbury stem from a probability conundrum that is sometimes called the Birthday Coincidence. The answer and explanation developed here is based on the treatment given to this puzzle in Richard Dawkins's book *Unweaving the Rainbow*.

The best way to work out how likely it is that two people among 23 in a room will share a birthday is to work out how likely it is that they don't, and then to subtract the answer from one. This is how you do the calculation.

Let's start by picking a person in the room randomly — maybe the archbishop. Obviously if there's just one person then the probability that there is no match is one — there is nobody with whom to match. Okay, so we'll bring a second person into the equation. What's the chance that this person, let's call her Betty, will share a birthday with the archbishop. Well, the archbishop only has one birthday, so it's one in 365, which means the chance that there is no match is 364/365. (Remember here that we're ignoring leap years.)

Let's add a third person, Charlie. What's the chance that he doesn't share a birthday with either the archbishop or Betty? This is a bit more complicated to calculate, but not too much of a problem. The key is to recall that we're calculating the odds that people do not share the same birthday. This means that we know that the archbishop and Betty have different birthdays, which in turn means that Charlie can't share a birthday with both of them. Therefore, the chance that Charlie will share a birthday with the archbishop is one in 365 and the chance he will share a birthday with Betty is one in 365. Turn this around, and you get a 363/365 chance that he will share a birthday with neither of them.

At this stage, we know that the odds that the archbishop doesn't share a birthday with Betty are 364/365, and the odds that Charlie doesn't share a birthday with either of them is 363/365 (given that the archbishop and Betty don't share a birthday with each other). To get the probability that both these things are true - that nobody shares a birthday with anybody else - we need to multiply these odds together. Thus, we get 364/365 363/365 as the odds that nobody among these three people shares a birthday with anybody else.

To add a fourth person, let's call her Danielle, you do exactly the same thing again. The odds that she won't share a birthday with any one of the archbishop, Betty and Charlie is 362/365, which you then add to the overall calculation to determine the odds that nobody will share a birthday with anybody else - 364/365 363/365 362/365.

That's it really. To get to the final answer you need to repeat this process for all 23 people in the room. The final multiplication sum, which gives you the probability that nobody in the room will share the same birthday, ends with 343/365. If you do the calculation, you'll find the answer is 0.49. Remember, this is the likelihood that there will be no shared birthdays in the room. Turn this around — that is, subtract the answer from one — and you have a slightly greater than 50/50 chance that somebody in the room will share a birthday with somebody else.

As Clare found out to her cost, the mix-up at the birthday soiree really wasn't that unlikely. If there are 23 people in a room, then half of the time two of them will share a birthday. So if you ever find yourself in a crowded room crouching in a giant cake waiting to leap out upon some unsuspecting birthday boy or girl, best check you're not about to pounce on the Archbishop of Canterbury.

IS THE CAT DEAD OR ALIVE?

(see pages 42-43)

The scenario depicted here is a version of a famous thought experiment proposed by physicist Erwin Schrödinger in 1935, which aims to demonstrate the shortcomings of an approach to quantum mechanics known as the Copenhagen interpretation. The issues at stake here are complex, so it is necessary to approach them in stages.

Perhaps the best starting point is to consider what's involved in tossing a coin. If you toss a coin, immediately cover it with your hand, and ask a friend to guess whether it's landed on heads or tails, two things are true about the situation: the first is that your friend doesn't know whether the coin is showing head or tails; the second is that there is a fact of the matter about it — whether the coin is showing head or tails is not in any sense dependent upon your friend's knowledge of the situation.

The key point is that the quantum world is not like this: at the quantum level, the state of affairs that exists in the world is integrally bound up with acts of observation and measurement. So, for example, if you look at an electron and determine that its spin is up, it does not follow that its spin was up before you looked at it. In fact, according to the Copenhagen interpretation, until you looked its spin was both up and down at the same time — it existed in a state of quantum superposition — and it took on a determinate direction only at the point at which it was observed.

The Schrödinger's cat thought experiment is designed to show that the consequences of this idea — that the state of affairs in the quantum world is dependent upon acts of observation and measurement — are absurd.

At the level of single atoms, radioactive decay is a stochastic process — that is, it is impossible to predict. The only way an observer can know whether the atom in the Schrödinger's cat

scenario has decayed is by smashing the box open to see whether the cat is still alive. However, until the box is smashed open, quantum superposition means the atom exists simultaneously in all its states, including decayed and not decayed, which seems to require the cat to be both dead and alive at the same time. This is obviously an absurd conclusion.

It is important to be clear that the point of this thought experiment is not to show that there can be such things as alive-dead cats. Rather, it is to show how the Copenhagen interpretation of quantum mechanics runs into choppy waters when applied to the world of everyday objects. Therefore, the correct answer to the question about Purccini's state of health is not that he is simultaneously both dead and alive, but rather that it is not possible to know whether he's alive until you look inside the box.

HOW OLD IS EINSTEIN'S TWIN?

(see pages 44-45)

The first thing to get straight when dealing with the complexities of special relativity, and the question of Egbert's age, is that time dilation is a real, empirically verified, phenomenon. It was decisively confirmed in 1971 by Joseph C. Hafele and Richard E. Keating, who boarded a civilian aircraft with four atomic clocks and flew twice around the world, first eastward, and then westward. Afterwards, the clocks were found to be discrepant in precisely the way predicted by Einstein's theory of relativity.

The effects of time dilation are very small at the sorts of speeds humans are currently able to reach. Keating estimated that if you spent your entire life flying around the world on a commercial airliner, you'd end up roughly one ten thousandth of a second younger than somebody who had stayed put. However, as you begin to approach the speed of light the effects become much greater, and the figures given in the Egbert and Albert example are accurate: if, from the perspective of somebody observing from Earth, you travel for ten years at 80% of the speed of light, you will return having aged only six years, and the clocks in your spaceship will confirm that this is how long you have been away.

The consequences of time dilation are undoubtedly a little weird, but there is nothing controversial about the phenomenon itself. The so-called twin paradox arises only when you combine time dilation with the fact that both twins can argue that it is the other twin who is moving, and who will therefore age less during the period of travel.

It is worth stating up front that the twin paradox is not a genuine paradox, but rather the consequence of a naïve understanding of the phenomenon of time dilation. In fact, the events it describes are perfectly explicable within the framework of Einstein's theory of relativity. It is possible to show how the paradox fails in a number of different ways, and there is some dispute about the most effective approach, but we're going to focus here on the significance of the effect of gravity upon time.

Paradox Resolved

Albert Einstein produced not one, but two, separate theories of relativity: special relativity, which has been the source of our reflections on the phenomenon of time dilation, in 1905; and general relativity, one of the two foundational pillars of modern physics, in 1915. The significance of general relativity for our purposes is that it describes how gravity causes time to run more slowly. A clock that is close to a massive body will be slower than a clock that is further away from the same body.

As with relative velocity time dilation, there is plenty of empirical evidence confirming the existence of gravitational time dilation. For example, in 2010, a research team at the National Institute of Standards and Technology led by Dr. James Chin-Wen Chou found that raising an atomic clock by just 30 centimeters caused it to run more slowly. Not by much, of course, a mere 90 billionths of a second over an average lifetime, but enough to

confirm the effects of a decrease in gravity.

This phenomenon is interesting and important in its own right — its effects are routinely taken into account by the GPS satellites that circle the Earth — but to understand its relevance to the twin paradox it is necessary to say something about what Einstein called the Principle of Equivalence, which holds that a "system in uniform acceleration is equivalent to a system at rest immersed in a uniform gravitational field."

This means, in effect, that acceleration mimics gravity. We already have a sense of this, of course, when we talk of g-forces, and the experience of being pushed back into our seat as an airplane or car accelerates, or as we ride a rollercoaster. The crucial point here is that just as gravity slows down time, so does acceleration, and it is this fact that provides an answer to the twin paradox.

The twin on the spaceship — Egbert, in our example — is subject to acceleration and deceleration. To put this more technically, he exists in an accelerated (non-inertial) frame of reference. The Earthbound twin, on the other hand, is not: he — Albert — exists in an inertial frame of constant velocity. This means that Egbert will age more slowly than his twin brother during the phases of acceleration and deceleration, and will return to Earth the younger of the two. So Egbert was right: at the end of his little jaunt, he will be the more youthful of the two twins.

The paradox of the twins gets going because it naively treats the perspectives of the twins as being symmetrical. But they are not: only the twin in the spaceship performs a U-turn in order to return to Earth, and only this twin undergoes the effects of acceleration and deceleration during this phase. The reason Egbert ages more slowly than Albert is because only he is immersed in the gravitational field resulting as a consequence of the change in velocity that occurs as the spaceship turns around to head back to Earth.

ARE THERE SPOOKY ACTIONS AT A DISTANCE?

(see pages 46-48)

The idea that particles can communicate information across unfathomably large distances at faster than the speed of light seems absurd. Einstein himself declared that he could not believe in quantum mechanics because "the theory cannot be reconciled with the idea that physics should represent a reality in time and space, free from spooky actions at a distance".

Einstein thought there must be "hidden variables", some underlying framework of laws, that would allow for deterministic explanations of quantum phenomena. Niels Bohr, one of the luminaries of the new quantum mechanics, disagreed, rejecting his friend Einstein's contention that "no reasonable definition of reality" would allow ghostly action at a distance.

Einstein's thought experiment had no definitive resolution during his lifetime. There was no way of deciding between his "hidden variables" approach and QM's ghostly action at a distance. However, this began to change in the 1960s with the work of theoretical physicist John Bell.

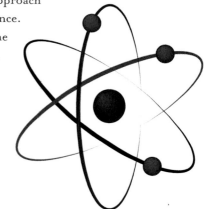

His inequality theorem has been called the "most profound discovery of science." It establishes an upper-limit on the synchronicity

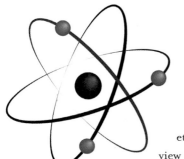

between isolated particles (precisely, isolated systems) such as those in Einstein's thought experiment, if it is true that these particles have observer-independent, determinate properties (e.g., spin, momentum, etc). If this limit is exceeded, then the QM view is right, and Einstein's hidden variables view is wrong.

The definitive experiment to test Bell's theorem was conducted by Alaine Aspect in 1982. It measured the extent to which pairs of photons cooperate as they hit polarizing material placed in their path. The result was in favour of the QM view. There was more synchronization between the particles than could occur given Einstein's contention that there must be hidden variables that determine the behavior of subatomic particles.

This is a striking result. It suggests we have to give up on the idea that there is an objective reality out there waiting to be revealed to us by the mechanism of systematic observation and measurement. Rather, the quantum world, at least, is a probabilistic realm of ghostly particles, indeterminacy, and action at a distance, that's only brought to reality by the act of observation.

The eagle-eyed reader might have noticed talk of information exchange at faster than light speeds. Unfortunately, quantum randomness rules out using entanglement to send meaningful messages at superluminal speeds. The British astrophysicist, John Gribbin, has a nice example that shows the problems involved.

Suppose you have two beams of entangled particles heading off in opposite directions towards a pair of detectors that are located a galaxy apart. A clever scientist at one of the detectors might find a way to mess around with the polarization of one of the beams, changing the spins of the particles in the beam. What is the person looking at the particles in the beam at the second detector going to see?

The answer is a random pattern of spins that is indistinguishable from any other random pattern. The scientist attempting to send the message cannot embed information in a pattern, because the pattern itself is the result of random quantum processes. The person at the second detector is going to see a pattern that's different from the pattern they would have seen had the clever scientist not messed around with the polarization of the first beam, but there's no information to be extracted there, because they have no way of knowing what the original pattern would have been. All they are seeing is a random pattern.

Quantum entanglement and its spooky action at a distance is perhaps the most shocking of QM's many oddities. Nevertheless, the evidence strongly suggests that we're going to have to learn to live with the shock. Einstein was almost certainly the greatest scientific mind of the 20th century. He was also almost certainly wrong about quantum mechanics.

WHEN IS A WAVE NOT A WAVE?

(see pages 49-51)

Thomas Young has been brought back to life to witness one of the most mind-boggling results in the whole of science. If you fire a single photon at a screen with two slits, and record the final destination of the photon on a photographic plate placed behind the screen, then common sense seems to dictate that the photon will travel through only one of the two slits, and then hit the screen somewhere broadly in line with the slit it traveled through (allowing for a certain amount of unpredictability) where it will create a single bright spot. If you do this over and over again, you should, given enough time, see two bright bands emerge that line up with the slits.

But actually you don't. Each photon does create a single bright spot, and repeating the process does build up a pattern, but it's an interference pattern (above), the kind that can only be generated when you have a separate wave coming from each of the two slits. The paradox here is obvious. A photon can surely only pass through one of the slits, but you get a pattern that requires overlapping waves that originate from both. The only possibility is that the photon has a particle aspect and a wave aspect.

In fact, this dual aspect is not exclusive to photons. The

experiment produces the same result if it's conducted with atoms, electrons, and other subatomic particles. Hence the concept of wave-particle duality is central to the branch of physics called quantum mechanics that deals with really, really small things.

So we have this notion that a photon has both wave and particle aspects. But the wave isn't a straightforward a physical thing; it's a mathematical construct, a function from which it is possible to derive the probabilities that govern the photon's ultimate destination. The wave function itself behaves like an actual wave, and splits as it emerges from the two slits, before eventually interfering with itself. It's this combined wave function that tells us that the photon is likely to be found in areas of constructive interference and not in areas of destructive interference.

If a wave function is probabilistic, the obvious question is what turns mere probability into the final determinate location of the photon (or other subatomic particle). The answer is the act of measurement. The moment when the position of the photon is determined is the moment when the wave function "collapses," and the photon then has an eigenstate of position, realizing one of the possibilities contained within the wave function, while all the others fall away. However, precisely how this occurs, what constitutes a measurement, and what happens to the other possibilities contained within the function is still a matter of considerable debate.

If you're a little confuddled by this stuff you're not alone. However, the equations that govern the field are well-established and allow for extremely accurate prediction. The real mystery lies in working out what quantum mechanics tells us about the nature of reality. And here things get very odd indeed. The German theoretical physicist Werner Heisenberg for example, suggested that what it actually reveals is that reality does not exist independently of our observations of it.

WHAT DID THE TELESCOPE REVEAL?

(see pages 52-53)

Alfred Russel Wallace's experiment to test the curvature of a six mile stretch of water produced the expected result. It would have been a bit of a worry had it not, given that the Earth actually is a sphere.

To recap, if the six miles of the river traced a convex shape, then the telescope would show the disc of the middle pole well above the black band that was affixed to the Old Bedford Bridge.

This is what was seen through the telescope on that Saturday morning early in March 1870, based on an illustration that appeared in *Field* newspaper later in the month.

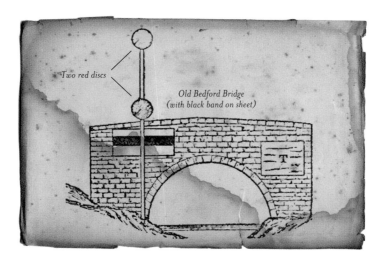

As you can see, both discs of the middle pole are above the black band. This is entirely as expected given that the Earth is a sphere and not a flat disc.

This should have been the end of the matter, but really it was only the beginning. John Hampden refused point blank to accept the result, claiming that the Old Bedford River was as flat as a billiard table, and you'd have to be mad drunk to think otherwise. He began a campaign of vilification and intimidation that quickly escalated into threats of violence, legal proceedings and ultimately his imprisonment.

Unsurprisingly, arguments over the substantive issue rumbled on. Part of the problem was that it is possible to do a version of Wallace's test and get a different result. For example, in May 1904, a photographer employed by a Lady Blount, a committed flat-earther, managed to get a photograph of a white sheet draped over the Old Bedford Bridge that he believed would not have been visible given a spherical Earth.

The likely explanation, assuming the experiment was set up correctly, is atmospheric refraction. Put simply, this refers to the way that light deviates from a straight line as a result of changes in air density. Refraction causes light to curve at least somewhat in line with the curvature of the Earth, which has the effect of "pulling up" objects that would otherwise be invisible below the horizon. This effect is more pronounced the closer you get to ground level.

Alfred Russel Wallace knew all about refraction, and compensated for its effect both in his calculations and in the placement of his telescope. But this didn't happen in the 1904 experiment, and refraction, together with other atmospheric effects such as temperature inversion, sufficiently explain the strange result.

The conflict between "globalists" and flat-earthers still rages to this day, mainly in the more esoteric corners of the internet. There is no doubt that the Earth is spherical, a fact that has been known for nearly 2500 years, but unfortunately this truth hasn't been quite enough to put the arguments to bed once and for all.

WHAT WILL THE CROCODILE DO?

(see pages 56-57)

The offer that Cuddly Croc has made to Adeline Doom mimics what is known as "The Crocodile Dilemma," which was first discussed by the Ancient Greeks. It has the following form:

A crocodile promises to return a child he has stolen, if and only if the father is able to tell him truly whether the crocodile will return the child or not.

We don't need to trouble ourselves too much about what happens if the father states that the crocodile intends to return his child. If he's right about it, he gets his child back. If he isn't, he doesn't. However, things get interesting if the father states that the crocodile will not return the child. In this situation, there are two possible outcomes, both of which result in a paradox.

Let's assume the crocodile intends to return the child. This means the father has not correctly identified what is going to happen, which means the crocodile will keep his child. However, if the crocodile keeps his child, it means the father has correctly identified what is going to happen, so the crocodile should return his child.

Okay, so let's assume the opposite, that the crocodile intends to keep the child. This means the father has correctly identified what is going to happen, which means the crocodile will return his child as agreed. However, if the crocodile returns his child, it turns out the father has not correctly identified what is going to happen, which means the crocodile should keep his child.

Both these outcomes are paradoxical. In neither case is it possible to determine whether the crocodile should return the child or not.

It is worth noting, though, that the paradoxical nature of these outcomes is not necessarily to the advantage of the father. If the father predicts that the crocodile will not return the child, he will never have a cast iron case that supports the return of his child. As the nineteenth century philosopher John Walker put it: "It is easy to see how the crocodile might retort this dilemma, and prove that in either case he must keep the child."

Following this, it's not entirely clear how Adeline Doom should respond to Cuddly Croc's offer. She could take a chance that The Croc's newly found conscience will have led him to decide to return her son, and then make that prediction in the knowledge that if she's right, she'll get her son back. Or, if she suspects The Croc fully intends to secure a ransom payout, she might predict that he will not return her son, and then hope to bamboozle him with the paradoxical consequences of such a prediction.

But perhaps an entirely new counter-offer would be the best strategy. Adeline Doom has many fine and delicious turkeys living on her various farms. Perhaps Cuddly Croc would be willing to return her son in exchange for a turkey or six. Then his conscience will be salved, he won't go hungry, and everybody will be happy. Except for the turkeys, of course.

TO VOTE OR NOT TO VOTE?

(see pages 58-59)

Toby is struggling with what is known as the paradox of voting, which German philosopher Hegel describes as follows:

> As for popular suffrage, it may be further remarked that especially in large states it leads inevitably to electoral indifference, since the casting of a single vote is of no significance where there is a multitude of electors.

The puzzle is normally couched as a problem for what's called rational choice theory. Most people are willing to incur the costs of voting — time, effort, expense of traveling, etc. — despite knowing that the chance of their vote being pivotal is effectively zero. The rational person, it seems, given that there is no obvious benefit to voting, should stay at home, yet people vote in the millions.

There is no doubt this paradox is troubling if you think people's behavior is motivated by their own personal cost-benefit analysis. Happily, though, there are other ways of looking at behavior, and Toby should have pointed out to Hypathia that the reasons people cast a vote are multifaceted and complex.

A person might choose to vote out of a sense of civic duty, or because they want to participate in the democratic process, or to express an opinion, or to influence others. They might vote to honor those who have gone before them in the struggle for freedom, or because it's what their parents would have wanted them to do.

In this sense, the paradox is not too difficult to escape. People vote not because they think their vote will make a difference, but for all kinds of other perfectly normal reasons — indeed, the kinds of reasons that often motivate behavior more generally.

However, there are aspects of the puzzle that are harder to dissolve. For example, it's not clear if it's possible to hold a person morally responsible for their voting decisions. If an individual's vote leaves the world exactly as it is — in all the morally relevant respects — then to what extent can they be culpable for their vote? At the very least, you're going to have to do some fast talking if you want to convince people that an act that has predictably no effects can nevertheless be immoral.

Let's, for the sake of argument, grant that the election of Donald Trump was a very bad thing. In electoral terms, and ignoring the complication of the electoral college, this large harm was caused by the collective action of some 63 million people, which means the contribution of any one individual to this harm was so small as to be negligible — in the absence of their vote, the result would have been the same. Therefore, it follows that no individual caused morally relevant harm, which in turn means they are not blameworthy, and there is no requirement for them to vote differently, or even more carefully, in the future.

This aspect of the paradox of voting is genuinely difficult to escape. Toby should have had no difficulty in rebutting Hypathia's first set of objections, but would have found himself on much more sticky ground had she asked him to explain *why* it would be wrong for her to vote against his preferred candidate.

THE SAME OR NOT?

(see pages 60-61)

Methuselah Labs' slippery CEO, Bradley Armitage, has concocted a tall tale in an attempt to escape the long arm of the law. In essence, his claim is that taking a single neuron from the brain of Person B (hereafter Ethel) and implanting it into the brain of Person A (hereafter Agnes) could never make enough of a difference that it would be reasonable to say that Agnes now has Ethel's mind. Personal identity seems to be a vague concept without sharp boundaries. It is implausible to think there is a sharp cutoff point at which we'd be justified in saying a person's thoughts and memories had changed enough for them to be considered a different person.

 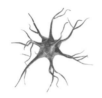 

However, without a sharp cutoff point, a single neuron will never be enough to bring about the transformation.

If this is so, it seems we have a contradiction. If it were possible to transplant neurons between brains, and to reconstruct neural networks so that they exactly mirrored the brain from which they originated, then at the end of the process we've described, Agnes and Ethel would insist that they had switched bodies, regardless of whether their brains were swapped over a single neuron at a time.

Armitage's sleight of hand is to suggest that this shows that Agnes could never become Ethel. However, that's not correct. In fact, what it shows is we have a contradiction between the claim that substituting a single neuron in a person's brain could never bring about the change from Ethel to Agnes; and what we know to be true about the world – that if Agnes has an identical copy of Ethel's brain she'll think she is Ethel (regardless of how she got there).

So how do we solve this contradiction? Well, it's not easy. This is a version of the Sorites Paradox, which is notoriously difficult to sort out. One possible strategy would be to argue that there would be a sharp cutoff point after which it would be accurate to say that Agnes is no longer the same person. That would solve the problem, because it would mean that the transplant of a single neuron within the long chain of neuronal substitution could push Agnes over line. But this is highly counterintuitive. Personal identity doesn't seem to be sharply defined – it's hard to imagine there's a clear dividing line after which we'd say, or Agnes would say, she was no longer the same person.

  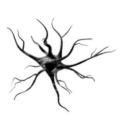 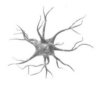

Happily, you don't have to sort out the Sorites paradox to put the sinister Armitage in his place. All you've got to do is to point out that we already know that what he says isn't possible is perfectly possible. That's what gives this situation its paradoxical character. Identical brains will manifest identical memories, dispositions, and so on. If Armitage has transformed Agnes's brain so that it is now a precise copy of Ethel's, then he has transplanted a mind between bodies. After all, if you ask Agnes who she is, and she answers truthfully, she'll tell you she's Ethel.

WILL THE TORTOISE START THE RACE?

(see pages 62-63)

The paradox of Telemachus the Tortoise and the gods who want to stop him in his tracks is a variation of a puzzle devised by philosopher José A. Bernadete in the 1960s, which itself was inspired by Zeno of Elea's classic paradoxes of motion.

The paradox has the following basic form. A god — Zeus in our scenario — waits in readiness to raise a barrier when Telemachus reaches the half mile point (in a one mile race). A second god (unknown to Zeus) intends to do the same thing at the quarter mile point. A third god at the eighth of a mile point. A fourth at a sixteenth of a mile, and so on, *ad infinitum*.

The paradox emerges because the set up suggests that Telemachus cannot move beyond the starting point — he could never move a distance sufficiently small so that no barrier would appear — and yet no god will ever get to raise a barrier, because there will always be an infinite number of gods earlier in the series waiting to raise a barrier. However, if no god is able to raise a barrier, then there is nothing to prevent Telemachus from moving beyond the starting point. So we have a contradiction. Telemachus both can and cannot get started.

The solution to this puzzle is contested, but the normal approach is to argue that there is a logical inconsistency in the way the problem is set up.

The gods are following a rule that in effect states: a barrier will appear at point p if and only if Telemachus reaches point p, which entails that no barrier has risen before point p.

Let's suppose Telemachus does get going — what happens next? There are two possibilities: either a barrier springs up or it doesn't. If it does, it means some god has raised a barrier despite the existence of an infinite number of barriers earlier in the series that should have been raised before point p. If a barrier doesn't appear, an infinite number of gods have refrained from raising a barrier despite Telemachus reaching their point *p*. Either way, as soon as Telemachus puts the system to test, the gods as a group do not and cannot succeed in executing their intentions.

This basic inconsistency has led the philosopher Stephen Yablo to suggest that all this paradox really shows is that either Telemachus stops before he gets started or the gods have to face up to the fact that they're not able to do what they promised to do.

However, this is perhaps not quite the end of the story. It still isn't entirely clear why any *particular* god cannot follow the rule as stipulated. Each god will know where Telemachus has got to on his journey, will know whether a barrier has already appeared, and will know whether the tedious tortoise has reached their own point p. So what exactly goes wrong? Why does incoherence result when we consider the gods as a collective? There is no clear answer to this question. It's one of the many puzzling ideas that comes up when studying the infinite, and the infinitely small.

WHAT HAPPENS NEXT?

(see pages 64-65)

Dr Forster has just sent a message two years into the past in the hope that it will prevent her beloved Mr Architect from seeking her out back then, a course of action that led to his demise in a road traffic accident. Let's assume he gets the message in time, and alters his behavior accordingly. What should we expect to happen in Dr Forster's present?

The answer is not immediately clear. If one accepts the premise that communication with the past is not ruled out as a matter of logic or by the laws of physics, then there doesn't seem to be any reason why we couldn't send warnings to our predecessors. It is implausible to suppose that no warning could ever result in an appropriate change in behavior, in which case it should be possible for people in the past to avoid dangers that otherwise would have done for them.

However, this immediately results in a paradox, a version of what is known as the Grandfather paradox. For example, let's suppose that Mr Architect doesn't set out to meet Dr Forster in her past. This means that the accident that prompted Dr Forster's warning would never have happened. If it didn't happen, then there was no accident for her to witness, and nothing to motivate her warning.

There are a number of ways to avoid this paradoxical conclusion. The first is to deny the possibility of time travel into the past (or communication with the past), not merely as a matter of physics, but also as a matter of logic. Time travel into the past necessarily results in contradiction and incoherence, so it's ruled out.

A less hard-nosed approach is to argue that what's prohibited isn't traveling to the past per se, but rather changing the past. According to this view, the past cannot be altered, because it has already happened. If we travel into the past, then our actions there form part of the causal background of the present.

What would this mean for our scenario? Well, nothing would change in Dr Forster's present. Mr Architect would still be killed in a road traffic accident, either because he didn't receive her warning or because her warning played some part in the causal chain that led to him being killed.

There is nothing inconsistent in this view, but it feels a little unsatisfactory, partly because it's suggestive of an infinite feedback loop. If we're able to learn about the past by visiting it, then why can't we use the information we gather there to adjust our attempt to alter it? The obvious response that this information gathering trip is also part of the causal story of our present just shifts the problem back a stage. Why can't we use our knowledge of this further fact to adjust our attempt to alter the past?

One thing is clear in all this, which is that in our scenario it really doesn't make sense to suppose that Mr Architect will magically appear at the lake house. However, it goes without saying that if this were the plot of a film then that's exactly what would happen.

HOW DO BEES FLY?

(see pages 68-69)

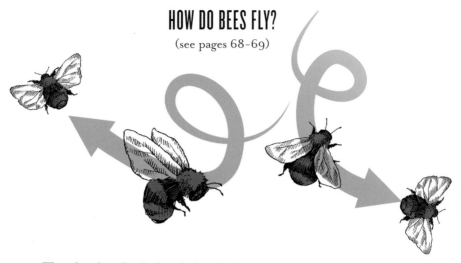

The idea that the flight of a bee defies the laws of physics is a well-known urban myth that likely dates back to the 1930s. Hopefully, it goes without saying that the humble bee actually poses no threat to our understanding of the physical world — if it did, then many more scientists would be working in the area of insect flight — but it is true that scientists have until recently had a great deal of trouble explicating the aerodynamic forces in play that keep bees, and other insects, in the air.

Part of the trouble is that a bee moves its wings at mind-boggling speed, nearly 250 beats a second. It also flies in a completely different way to an aircraft, which stays aloft because of the correct balance between the four aerodynamic forces of flight — lift, weight, thrust and drag — all of which remain in a steady state to keep an aircraft airborne.

If you pretend that a bee is a tiny plane, then it can't fly. It doesn't generate enough lift to counteract its weight. Even if you take into account that a bee doesn't fly like a plane, that it beats its wings, that the velocity of its wings changes during their motion, and so on, you still can't explain its flight using the principles that work for aircraft. Instead, you need to look at dynamic forces and unsteady airflows*.

Scientists, such as Caltech's Michael Dickinson and his colleagues, use a variety of techniques to investigate the aerodynamic effects that keep bees, and other insects, aloft. They build models, use slow motion video, take pictures, and conduct wind tunnel experiments. This research means we now have a very good idea of how bees manage to fly.

It turns out that a bee doesn't flap its wings up and down. Rather, its wings move in a rotating, back-and-forth pattern that traces a narrow oval. This complex pattern has the effect of creating tiny air vortices — mini tornados — around the wings that provide extra lift. A bee shares this ability in common with other flying insects, but there are some differences in how a bee flies that are instructive.

A bee compensates for its relatively stubby wings by beating them much faster than would be expected given its size, but across a shorter arc than is normal for flying insects. It also has the ability to respond to the demands of a heavier workload by increasing the arc of each beat while maintaining the same rate of beat.

According to Dickinson, bees exploit some of the most exotic flight mechanisms that are available to insects. It is hardly surprising, then, that it took scientists a long time to figure out exactly how bees manage to take to the air. But it is now surely time for the old myth that bees defy the laws of physics when they fly to be cast into the trash can of history.

In a strange kind of way, Eduardo got everything entirely backwards. If you compare the evolutionary success of flying insects to that of eagles, there's only one winner, and it isn't the species with the large beak and an attitude problem.

* See Michael Dickinson, "Solving the Mystery of Insect Flight," *Scientific American*, 2001.

OVER THE LIMIT?

(see pages 70-71)

To understand the pitfalls of Nicholas Trink's drug screening program, we must examine the difference between a false positive and a false negative. Trink believes his drug screening program is a reliable tool for weeding out drug users because it never produces a false negative — it will never incorrectly identify a person who has used drugs within the preceding 48 hours as somebody who hasn't.

The problem is that this doesn't rule out the possibility of a false positive — the possibility of incorrectly identifying somebody as a drug user when in fact they haven't used drugs. A screening program that identifies everybody as a drug user will never produce a false negative — it will never incorrectly identify a user as a non-user — but it will be overrun by false positives for the obvious reason that it will identify every non-user as a drug user.

Trink's screening program isn't quite this hopeless in its ability to distinguish between drug users and non-users, but if we do the math we'll find that it is actually fairly hopeless. The question is, what is the probability that somebody has used an illegal substance given that they test positive on the blood test?

To work this out we need to use something called Bayes' Theorem. We won't look at how this works in any detail; suffice it to say that it allows for probability calculations given certain kinds of prior knowledge. Let's plug our figures into the theorem, and see what result it gives. For our scenario, the calculation is fairly straightforward. This is the formula:

$$P(A/B) = \frac{P(B/A) \times P(A)}{P(B)}$$

P(A) is the probability of somebody having used an illegal substance within the preceding 48 hours

P(B) is the probability of getting a positive test result

P(A|B) is the probability of A given that B is true — the probability that somebody has used an illegal substance given that they test positive on the blood test (i.e., what we want to find out)

P(B|A) is the probability of B given that A is true — the probability of a positive test result given that a person has used an illegal substance within the preceding 48 hours

In our scenario:

P(A) = 0.001; 1 in a 1000 people have used an illegal substance within the preceding 48 hours

P(B) is 0.08; the screening program consistently shows an 8% positivity rate

P(B|A) = 1; everybody who has used an illegal substance within the preceding 48 hours will test positive

If we plug those figures into the formula we get:

$$P(A/B) = \frac{1 \; x \; 0.001}{0.08}$$

Do the math, and it turns out that the probability that somebody has used an illegal substance *given* a positive test result is 1.25%. Put simply, only 1.25% of people who test positive on Trink's test have actually used a drug within the preceding 48 hours.

Therefore, it seems that the denizens of Saint Pier's Facebook group are correct. There are an awful lot of people currently self-isolating on that rocky outcrop who have never taken a drug, and they don't even have the benefit of an altered consciousness to get them through the ordeal.

WILL SHE BE FASTER?

(see pages 72-73)

Midge's plan to gain an edge in her attempt to break Boastful Barry's mile world record for bees might sound ingenious, but it is, of course, a total nonsense. A few examples will suffice to make this fact intuitively clear.

Imagine you're on a train speeding along at 150 miles per hour, and you jump straight up into the air. What's going to happen? Hopefully it's obvious that you're going to land at the same point from which you took off. In this case, you have zero velocity relative to the train, even though relative to the world outside, you are traveling at 150 mph (along with the train and everything else inside it).

Let's suppose you walk down the carriages of the train at five miles per hour in the opposite direction to which it is traveling. What's going on here? Well, your velocity relative to the train is five miles per hour (as is the train's velocity relative to you), but relative to the outside world, you're still traveling at 145 mph in the opposite direction to which you're walking, a fact that you'd discover in dramatic fashion if you stepped off the train.

Now let's factor the spin speed of the Earth into this scenario, and see where that leaves us. Assume that the train is traveling at 150 miles per hour east to west on the equator, and inside the train you're traveling at 5 miles per hour west to east relative to the train.

This leaves us with the following set of velocities.

The Earth, at the equator, is spinning at 1000 miles per hour relative to its axis.

The train is traveling at 150 miles per hour relative to the Earth, and at only 850 miles per hour relative to the Earth's axis (thus, if you're on the train for enough time, you'll notice that it takes longer for your day to come to an end - moving against the spin of the Earth means moving more slowly than the Earth relative to the Earth's axis).

You are traveling at 5 miles per hour relative to the train, 145 miles per hour relative to the Earth, and 855 miles per hour relative to the Earth's axis (thus, if you were able to walk on indefinitely, you'd notice that your day would be a little shorter than that of a passenger who stayed still).

So what does all this mean for Midge's attempt to use the Earth's rotation to help her gain an edge in her attempt to break the mile world record?

Unfortunately, it means she's going to get no help at all. In the scenario we're envisaging, her speed over the mile is relative only to the distance over the ground. She's already moving, along with the ground, at 1000 miles per hour relative to the Earth's axis (if Scranton zoo were located on the equator). If she takes a minute to cover the mile, then her velocity relative to the ground is 60 miles per hour. In terms of the world record, it really doesn't matter that for the one minute duration of her attempt, she was only moving at 940 miles per hour relative to the Earth's axis.

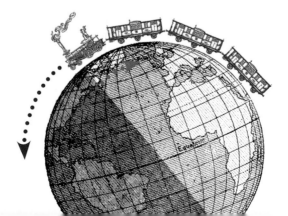

WHO WROTE THE MUSIC?

(see pages 74-75)

SEPT. 09.25.1966

Richard Ramme, Antonio Vivaldi's number one fan, has traveled back in time to meet the great baroque composer, and has gifted him a transcript of Vivaldi's own celebrated composition, *The Four Seasons*. Unfortunately, and unbeknownst to Ramme, Vivaldi had not yet written the four violin concerti that make up *The Four Seasons*, leaving the composer perplexed as to the origin of works that bear his own unmistakable stylistic imprint.

The obvious question here is who actually wrote *The Four Seasons*?

The only plausible answer is that nobody did. The four violin concerti have jumped into existence out of nothing. The events depicted in our scenario constitute what is known as a causal loop. Vivaldi is only able to put his name to *The Four Seasons* because Ramme delivers a copy of the music to him. But Ramme can't listen to a performance of *The Four Seasons*, transcribe it, and deliver it to Vivaldi unless the work already exists with its own particular history. Put simply, each event depends on the other.

This kind of time travel puzzle is known as the "bootstrap" paradox (after Robert Heinlein's novella, *By His Bootstraps*). It is often presented in the form of a story about somebody traveling into the past in order to share the secrets of a particular technology, where the ability to do so is predicated upon the technology having been discovered precisely as a result of the time traveler's actions.

One interesting point about the bootstrap paradox is that, strictly speaking, and unlike the Grandfather paradox (see "What

Happens Next?", p. 64), it doesn't result in incoherence and contradiction. You don't assert a contradiction if you posit the existence of self-causing or self-existing objects (the cosmological proof of God, for example, relies on the idea of an uncaused cause). It's just that we have no reason to believe such things routinely exist in our world. Objects, certainly complex objects, don't tend to pop into existence fully formed, so we find it almost impossible to believe that *The Four Seasons* could be self-causing.

But our intuitions about this are not sufficient to rule out the possibility of self-causing objects and uncaused events. The puzzle here is genuine. In a universe where time travel exists, what exactly stops somebody from traveling back to its version of 1680, for example, and letting Sir Isaac Newton in on the secret of his laws of motion? It's not clear that there is anything to stop it, certainly not the existence of unexplained objects and events. If so, the bootstrap paradox, in and of itself, cannot demonstrate the impossibility of time travel.

This must be good news for Richard Ramme, Antonio Vivaldi's greatest admirer. If he lives in a universe where backward time travel is allowed, then it's entirely possible that he's an integral part of the causal story of the existence of *The Four Seasons*. It's a shame, of course, that it also means that nobody actually composed the four violin concerti he reveres above all other works of art.

WHERE ARE ALL THE ALIENS?

(see pages 76-77)

The Italian-born American physicist, Enrico Fermi, came up with the paradox that now bears his name at a luncheon in the summer of 1950. There are competing accounts of exactly how it happened, but the gist of the story is that Fermi and his colleagues had agreed that intelligent life likely exists on other planets within our galaxy, which prompted Fermi to ask, "But where is everybody?".

Although Fermi never developed a formal account of his paradox, the question he posed remains very much alive to this day. The basic claim underpinning the paradox is that the vast age of the Milky Way means that however you do the math, whatever baseline figures you use to work out how long it will take for technologically sophisticated civilizations to emerge, how fast extraterrestrial spacecraft will be able to travel, how quickly colonized planets will develop their own colonial ambitions, you'll still easily have enough time for the colonization of the galaxy.

But if that's true, where are all the aliens?

Probably the first systematic attempt to answer this question was provided by astrophysicist Michael H. Hart in his 1975 article, "An Explanation for the Absence of Extraterrestrials on Earth." Hart states that the obvious answer is that aliens don't exist, but he identifies four kinds of alternative explanations for their absence.

Physical: *Physical explanations hold that there are insurmountable problems confronting any species that wishes to travel between star systems, whether this be a matter of the vast distances involved, engineering difficulties, or the biological challenges of interplanetary travel.*
Sociological: *Sociological explanations hypothesize about the structure of*

extraterrestrial civilizations, and their politics, interests, motivations, history, and so on, to explain the absence of aliens on Earth. The standard idea is that aliens could visit Earth, but choose not to do so.

Temporal: *Temporal explanations rely on the claim that there's not been enough time for aliens to visit the Earth.*

They Were Here: *These explanations hold that aliens have visited Earth in the past. It's just that they are not here now. (The number of reputable scientists who take this possibility seriously is vanishingly small.)*

Hart finds all these explanations lacking. For instance, he accepts that it is not a trivial feat to travel between star systems, but denies that the difficulties are insurmountable. Such journeys might be undertaken by artificial lifeforms, for example, meaning time constraints are not a large hindrance. Or they might be planned and completed across more than one generation of travelers.

Similarly, while it might be true that in some alien societies, sociological factors work against space colonization, if intelligent life is remotely common in the galaxy, it's hard to believe that these factors would intervene in all cases. Sophisticated civilizations might tend to blow themselves to bits before they reach technological maturity, but it's unlikely they would all do so.

Hart's conclusion will not please Badger Blight at all. Hart argues that in light of the fact there are no aliens on Earth, it's highly unlikely the galaxy is populated by thousands of intelligent civilizations. In fact, as Danielle Scullery asserts, there could be no such civilizations. Two things follow from this. First, the search for extraterrestrial intelligence (SETI) is probably a waste of time. Second, in the future, it's probably our own descendants who will occupy most of the Milky Way's inhabitable planets.

INDEX

A

Age Problems 22–23, 94–95
aliens paradox 76–77, 140–141
Are There Spooky Actions at a
     Distance? 49–51, 115–117
Are We Living in a Simulation?
     30–31, 100–101
Are You the New Sherlock Holmes?
     12–13, 82–83
Aspect, Alaine 116

B

Bayes' Theorem 134
bee flight 68–69, 132–133
belief bias 83
Bell, John 115–116
Bernadete, José A. 128
Birthday Coincidence 38–39,
     108–109
Bohr, Niels 41, 115
bootstrap paradox 74–75, 138–139
Bostrom, Nick 100

C

Carroll, Lewis 9, 91
Who Lost What? 20–21, 91–93
classic puzzles and conundrums 9
Are You the New Sherlock Holmes?
     12–13, 82–83
How Many Cookies Will Be Left?
     18–19, 88–90
How Wide is the Lake? 16–17,
     86–87
Married or Not? 10–11, 80–81
Switch or Stick? 14–15, 84–85
Who Lost What? 20–21, 91–93
Copenhagen interpretation 42–43,
     110–111
Crocodile Dilemma 56–57,
     122–123

D

Dawkins, Richard 108
Dickinson, Michael 133

E

Einstein, Albert 6, 41, 67
Are There Spooky Actions at a
     Distance? 49–51, 115–117
How Old Is Einstein's Twin?
     44–45, 112–114
Evans, Jonathan 83

F

Fermi paradox 76–77, 140–141
Feynman, Richard 41

G

Gardner, Martin 96
Gott, J. Richard III 98
Grandfather paradox 6, 64–65,
     130–131
Gribbin, John 116–117

H

Hafele, Joseph C. 112
Hampden, John 52–53, 120–121
Hart, Michael H. 140–141
Hegel, G.W.F. 124
How Do Bees Fly? 68–69, 132–133
How Long Will Their Relationship
     Last? 28–29, 98–99
How Many Cookies Will Be Left?
     18–19, 88–90
How Many Possible Worlds? 34–35,
     104–105
How Old Is Einstein's Twin?
     44–45, 112–114
How Old is He? 22–23, 94–95
How Wide is the Lake? 16–17,
     86–87

I

Is the Cat Dead or Alive? 42–43,
     110–111

K

Keating, Richard E. 112

L

Levesque, Hector 80
Lewis, David 104

logic puzzles:
Are You the New Sherlock Holmes?
    12–13, 82–83
Married or Not? 10–11, 80–81
Switch or Stick? 14–15, 84–85
Loyd, Sam 86

M

Married or Not? 10–11, 80–81
modal realism 34–35, 104–105
Monty Hall puzzles 14–15, 84–85
Morris, Dan 25

N

natural science puzzles 67
How Do Bees Fly? 68–69, 132–133
Over the Limit? 70–71, 134–135
Where Are All the Aliens? 76–77,
    140–141
Who Wrote the Music? 74–75,
    138–139
Will She Be Faster? 72–73, 136–137

O

Over the Limit? 70–71, 134–135

P

paradoxical puzzles 55
How Old Is Einstein's Twin?
    44–45, 112–114
The Same Or Not? 60–61, 126–127
To Vote or Not to Vote? 58–59,
    124–125
What Happens Next? 64–65,
    130–131
What Will the Crocodile Do?
    56–57, 122–123
Where Are All the Aliens? 76–77,
    140–141
Who Wrote the Music? 74–75,
    138–139
Will the Tortoise Start the Race?
    62–63, 128–129
prediction puzzles:
How Long Will Their Relationship
    Last? 28–29, 98–99
What Will the Crocodile Do?
    56–57, 122–123

Principle of Equivalence 114
probability puzzles 25
Are We Living in a Simulation?
    30–31, 100–101
How Long Will Their Relationship
    Last? 28–29, 98–99
How Many Possible Worlds? 34–35,
    104–105
What's the Real Benefit? 36–37,
    106–107
Which Die Should She Choose?
    32–33, 102–103
Who Will be Executed, Who Will be
    Pardoned? 26–27, 96–97
Whose Birthday Is It Anyway?
    38–39, 108–109

Q

quantum entanglement 49–51,
    115–117
quantum mechanics 6, 41
Are There Spooky Actions at a
    Distance? 49–51, 115–117
How Old Is Einstein's Twin?
    44–45, 112–114
Is the Cat Dead or Alive? 42–43,
    110–111
What Did the Telescope Reveal?
    52–53, 120–121
When Is a Wave Not a Wave? 46–48,
    118–119

R

refraction 121
relativity theory 6, 44–45, 112–114
risk, absolute and relative 36–37,
    106–107

S

Same Or Not? 60–61, 126–127
Schrödinger's cat 42–43, 110–111
simulation arguments:
Are There Spooky Actions at a
    Distance? 49–51, 115–117
Are We Living in a Simulation?
    30–31, 100–101
Sorites paradox 60–61, 126–127
Stanovich, Keith 81